CEREBRAL MONITORING
IN THE OPERATING ROOM AND THE INTENSIVE CARE UNIT

DEVELOPMENTS IN
CRITICAL CARE MEDICINE AND ANESTHESIOLOGY

Volume 22

Cerebral Monitoring in the Operating Room and the Intensive Care Unit

ENNO FREYE, M.D.
Department of Vascular Surgery and Renal Transplantation
Heinrich Heine University of Düsseldorf
and
Department of Central Diagnostics,
Rheinische Landes- und Hochschulklinik in the University Clinics of Essen
F.R.G.

Kluwer Academic Publishers
Dordrecht / Boston / London

Library of Congress Cataloging-in-Publication Data

Freye, E. (Enno)
Cerebral monitoring in the operating room and the intensive care unit / by Enno Freye.
p. cm. -- (Developments in critical care medicine and anaesthesiology : 22)
Includes bibliographical references.
ISBN 0-7923-0439-X (U.S.)
1. Electroencephalography. 2. Evoked potentials (Electrophysiology). 3. Patient monitoring.
4. Therapeutics, Surgical. 5. Critical care medicine. I. Title. II. Series.
[DNLM: 1. Critical Care. 2. Electroencephalography. 3. Evoked Potentials. 4. Intraoperative Care.
5. Monitoring, Physiologic. WL 150 F893c]
RC386.6.E43F74 1990
616.8'047547--dc20
DNLM/DLC
for Library of Congress 89-19791
CIP

ISBN 0-7923-0439-X

Published by Kluwer Academic Publishers,
P.O. Box 17, 3300 AA Dordrecht, The Netherlands.

Kluwer Academic Publishers incorporates
the publishing programmes of
Martinus Nijhoff, Dr W. Junk, D. Reidel, and MTP Press.

Sold and distributed in the U.S.A. and Canada
by Kluwer Academic Publishers,
101 Philip Drive, Norwell, MA 02061, U.S.A.

In all other countries, sold and distributed
by Kluwer Academic Publishers,
P.O. Box 322, 3300 AH Dordrecht, The Netherlands.

Printed on acid-free paper

Printed in the Netherlands

Contents

Preface

In spite of today's increasing body of knowledge in regard to central nervous function and/or the mode of action of centrally active compounds, little is done to monitor those patients which are at risk of cerebral lesions either in the OR or in the ICU. Due to the inconsistency of reports regarding the application and the benefits computerized EEG and/or evoked potential monitoring will bring to the clinician, physicians still are reluctant to get involved with a technique, which they think, will have little or no effect on the outcome of a patients well being. However, due to the development in computer technology, data acquisition and comprehension, it now is possible to monitor such a viable organ as the Central Nervous System (CNS) on a routine base without being a specialist in neurology or electroencephalography. Thus, the book is intended to guide the clinician to use EEG and evoked potential monitoring in a day to day situation, without going too deep into technical details. As an improvement of cerebral care is needed, various representative cases underline the interpretation of EEG power spectra and evoked potential changes in regard to the underlying clinical situation. It is hoped that this book will serve as a guide to anyone who considers cerebral monitoring a necessity in today's patient care. This may be the anesthesiologist, the intensive care therapist, the nurse anesthetist as well as the medical personnel in the ICU setting. The aim therefore is to give an idea of what cerebral monitoring can do, what its limitations are, and how to interpret the data in the light of the other physiological variables. Aside from a short description of the devices which are on the market, emphasis is placed upon the technique of electrode placement, data interpretation, and the techniques of how to avoid the recording and processing of data which is contaminated with electrical noise. In the light of data being derived by cerebral monitoring, the data derived from other monitoring systems plus the clinical impression, a diagnosis in regard to cerebral function can be made on more reliable grounds. Suitable to use as an introduction or as a reference, *Cerebral monitoring in the OR and the ICU* presents

- complete coverage of all aspects of EEG power spectra and evoked potential monitoring
- the major features of EEG power spectra and evoked potential machinery including amplifiers, filters, and microprocessors

- suggestions for assessing and improving signal quality, reducing noise and artifacts encountered during monitoring in the OR and the ICU
- examples of use of EEG power spectra and evoked potential monitoring especially related to anesthetic agents, coma, head trauma, and ischemic events.

In addition, coverage of intraoperative monitoring of various evoked potential modalities with a list of test parameters of typical latency and amplitude criteria is provided.
As an introduction, recommendations are given for in-depth coverage of practical organization to guide the beginner.

Many thanks to Mrs. Karen Schrecker (San Diego, USA) and Boris Neruda, M.D. (Düsseldorf, FRG), for proof reading and stylistic corrections.

Düsseldorf and Essen, 1990 Prof. Dr. Enno Freye, M.D.

1. Introduction

1.1 Rationale for the use of cerebral monitoring in the OR and the ICU

More than any other part of the entire organism the central nervous system is affected profoundly by anesthetics. Additionally, episodes of malperfusion or of hypoxemia during or after anesthesia are seldom detected shortly after their initiation, as monitoring of the CNS is not a commonly used technique during anesthesia or in the intensive care ward.

Yet, the methods which are commonly employed to monitor the CNS have *not* changed considerably, although nowadays various techniques are available for the clinician to gain further insight into cerebral function. Despite the fact that the target organ of anesthetics is the CNS, the various planes of anesthesia commonly are evaluated by means of cardiovascular parameters such as blood pressure, heart rate, and the respiratory rate, or movements of the limbs in response to nociceptive stimulation. However, with the widespread use of neuromuscular blocking agents, some of these simple signs of anesthetic depth are not available any more. Thus, unsuspected awareness during anesthesia which often is also associated with psychic stress and physical pain, is not too uncommon [1, 2].

Apart from such drawbacks, the anesthetic state itself limits the ability of the anesthesiologist or the surgeon to evaluate those parts of the CNS which may undergo such serious injuries as they are common during some neurosurgical, orthopedic and/or cardiovascular operations. Thus, the electroencephalographic activity of the brain transmitted to the electroencephalogram (EEG) represents a 'monitor' of the organ which is not only the target of anesthetic agents, but also the most *sensitive* to ischemia. Therefore, the major uses of the EEG have involved the estimation of anesthetic depth and the detection of ischemia.

In the Intensive Care Unit (ICU) the pharmacologically depressed state of the patient's consciousness often hampers neurologic assessment in situations when information about the functional integrity of neural structures is most needed, such as to guide treatment or to predict outcome.

In order to circumvent these drawbacks in today's anesthesia and intensive care management, electrophysiological monitoring of the nervous system slowly is gaining widespread use in the OR and the ICU [3, 4].

Application of techniques for the *noninvasive monitoring* of the CNS activity and the functional integrity of neural structures in the operating room and the critical care unit used to be very complex and expensive. Thus, not many centers engaged in such activities which provided them with additional information to prevent neurologic injury [5, 6]. Due to the rapid development in computers, and computer-aided data acquisition and comprehension in the past, electrophysiological monitoring is gaining new interest. This is partly because neuromonitoring enables the clinician to gain new insights into the mode and the site of action of various centrally active compounds. But more so, computer application has simplified neuromonitoring to such a degree that little additional training is necessary to sufficiently derive electrophysiological data from patients under anesthesia and read graphs with sufficient ease which are computed by microprocessors and displayed on a screen.

Computers and microprocessors are much faster in data acquisition, transformation, and comprehension; thus, they are much more suitable to take over a good deal of work in electrophysiological monitoring. For instance, an experienced electroencephalographer would be needed to examine EEG readings which, when they are recorded, come in different shapes of waves while extracting certain features from these tracings. He would process the EEG waves in his mind using the data base to operate with, compare the data with past information and finally produce a result. Computer processing as it is used nowadays uses a similar approach. While examining the entire tracing, specific features are extracted and processed in a similar fashion. Just what features are extracted and how well they are presented determines the usefulness of computer-processed monitoring of central activity. Thus, in the past, various concepts in the CNS data processing and evaluation have been developed, some of which have broadened our view of the CNS and have given us additional insight into the reaction of the CNS to therapeutic manoeuvering. Some of these concepts have lived a short life as processing was not sophisticated enough and the display too complicated to gain any additional information [7, 8].

In the development of EEG recording, interest among clinicians has been intermittent burst of enthusiasm followed by periods of little or no activity. By contrast, use of the electrophysiology in neurological investigation developed rapidly and has maintained its importance as a diagnostic tool so that virtually every large hospital has its own EEG department. This, however, has not been taken up by the anesthesiologist and those clinicians working in the intensive care unit, as confusion had spread about the use of neurophysiologic monitoring in the OR and/or the ICU. The cause of this confusion stems largely from the complicated technique and the size of the device necessary to run an EEG analysis. Although the same signal may be under investigation, there are distinct differences in regard to interpretation. The EEG is very different, where a focus of abnormal electrical activity is undergoing investigation, compared to a normal patient who shows EEG changes due to an anesthetic drug or hypoxemia [9–13]. In both cases the signal comes from the same electrical activity of the CNS, the application, however, is very different. For the anesthesiologist or intensive care therapist who wishes to make intelligent use of the information provided by electrophysiological monitoring, the first requirement is to be able to recognize the standard signal associated with abnormal physiological changes. Therefore, a rational assessment can be made of the variations in clinical patterns induced by anesthetics and various comatose states. The basic information which is needed for this approach is set out below referring to the latest development in computerized EEG and evoked potential recording while examples in various clinical settings will be used to underline the potential advantages of electrophysiological monitoring.

1.2 Why monitor the brain

Until recently the EEG monitoring outside the diagnostic laboratory has met limited use despite its clear role in areas such as anesthesia, trauma, and intensive care treatment.

There are primarily three reasons why the clinician resented the fact of using the electroencephalographic reading in his daily work:

1. Laborious reading of the bulky tracing.
2. The size and the complexity of conventional EEG equipment.
3. The need for a trained physician to run the machine [14].

The advances in electronic technology, however, have overcome these difficulties aiming at a synthesis of the most significant aspects even at the cost of sacrificing details. One of the various procedures, the transformation from the time into the frequency domain and the presentation of the spectra in a quasi 3-dimensional fashion offered a comprehensive picture of the base EEG activity and of the distribution of the dominant frequencies and their revolution in time.

Practicing this method, it now is possible to calculate in real time the spectral power amounts of the EEG which are useful in a number of clinical settings [15]:

1. To provide continuous information of the onset, depth, and the duration of the anesthetic on cortical structures.
2. To observe arousal during anesthesia which could have been induced by surgical stimuli.
3. To anticipate the time of awakening and recovery from anesthesia.
4. To assess completeness of recovery from the anesthetic.
5. To confirm satisfactory cerebral perfusion during special types of operation such as cardiopulmonary by-pass, carotid endarterectomy, and during deliberate hypotension.
6. To provide a continuous and objective record of sufficient cerebral oxygenation.
7. To immediately assess the effects of hypoperfusion and/or hypoxemia on the brain and effectively take corrective actions.

As the EEG is a minute, brain-generated voltage differential between two arbitrary sites on the scalp, these potential changes represent only a small portion of the brain. It can be estimated that the actual EEG represents a sampling from no more than one third of all brain cells. As to the site of origin of cortical potential changes, it appears that the post-synaptic potential changes of the cortical pyramidal or Betz cells play an important part in generating the electrical changes (Fig. 1 [16]).

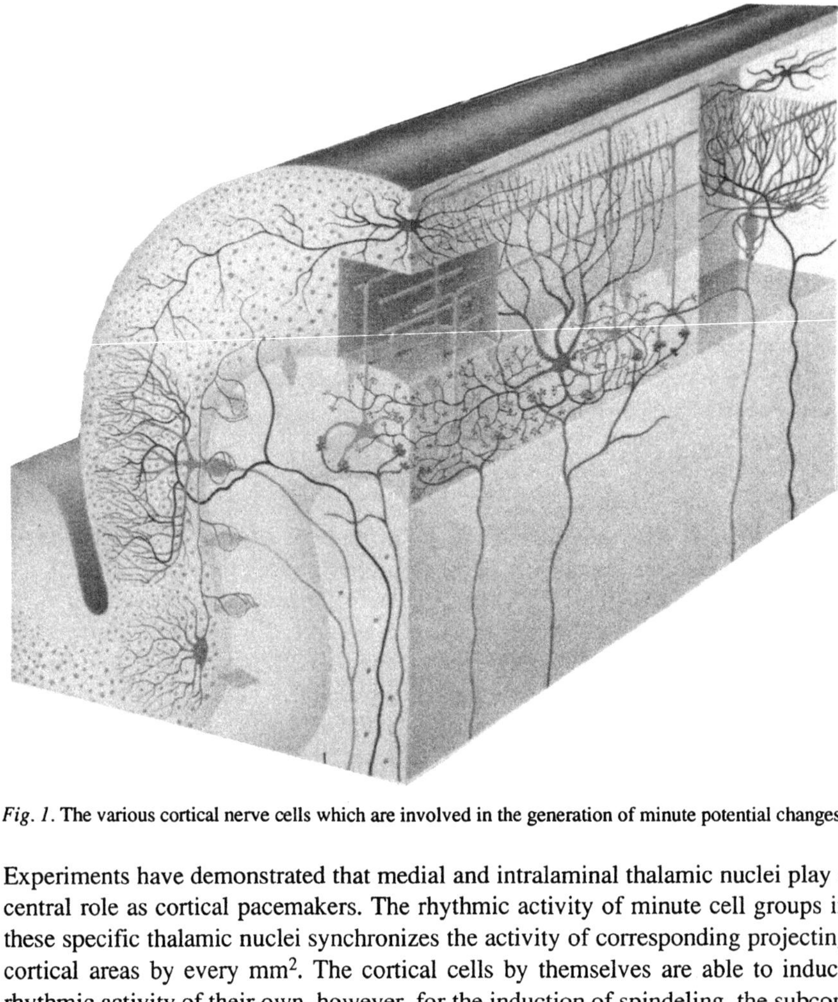

Fig. 1. The various cortical nerve cells which are involved in the generation of minute potential changes.

Experiments have demonstrated that medial and intralaminal thalamic nuclei play a central role as cortical pacemakers. The rhythmic activity of minute cell groups in these specific thalamic nuclei synchronizes the activity of corresponding projecting cortical areas by every mm^2. The cortical cells by themselves are able to induce rhythmic activity of their own, however, for the induction of spindeling, the subcortical pacemaker seems to be a prerequisite [17, 18].

1.3 The neurological approach to EEG interpretation

Patterns of the EEG are recognized by analyzing the signal with respect to its frequency, amplitude, and asymmetry:

(a) **Frequency** is the measure of the number of times a respective wave reoccurs in one second interval. Most of the EEG activity occurs between 1–30 cycles per second which equals the dimension *Hertz* (Hz). The frequencies of the EEG waves traditionally have been divided into four standard bands:

delta	=	0.5– 3 Hz
theta	=	4 – 7 Hz
alpha	=	8 –12 Hz
beta	=	13 –30 Hz (see example Fig. 2)

Fig. 2. Representative example of the various EEG frequencies appearing at different situations in life. Each band is associated with certain physiological states: beta is referred to as being awake and alert, alpha dominance is found in a relaxed and drowsy state, while delta and theta waves indicate pathology except during the different stages of sleep.

(b) The **amplitude** refers to the height of the EEG waves as measured in microvolts (μV) and categorized rather loosely in following subunits:

low amplitudes	=	less than 20 μV
medium amplitudes	=	20–50 μV
high amplitudes	=	more than 50 μV.

(c) The **symmetry**: generally the EEG of the two hemispheres is symmetrical. Anesthetic agents usually affect both hemispheres symmetrically. However, impediment of the blood flow to the brain, such as in carotid clamping, for instance, may result in asymmetry.

2. The principle of EEG recording using computerized power spectral analysis

2.1 Difference with conventional EEG recording

The EEG measurement and recording system, although analogous to that used for *ECG* measurement, varies in that the EEG machine:

a) has additional flexibility, and
b) is more susceptible to artifacts since the potential changes of the superficial, cortical layers are amplified by a factor of 100.

Thus, the ECG potential changes are measured in millivolt (mV), those of the EEG in microvolt (μV).

The recording instrumentation is of utmost importance as poor recording electrodes result in a poor signal, lots of artifacts, and difficult interpretation. For greatest accuracy it is necessary to use non-depolarizable electrodes which are capable of recording frequencies over a wide band pass.

Three types of electrodes are generally used in the operating room and the intensive care unit:

1. Carefully prepared silver or gold cup electrodes about 1 cm in diameter and coated with silver chloride. They are fixed to the scalp with *collodion*, an adhesive plaster, and are filled with saline gel similar to the ones used for ECG recording. The main purpose of this gel is to lower the impedance (Fig. 3).
2. Alternatively, sterilized stainless steel or platinum subdermal needle electrodes can be used. Their main advantage is that they may be quickly applied to the patient. However, they have a high *impedance* which makes them susceptible to electrostatic interference from other electrical apparatus nearby (Fig. 3).
3. For fast application on areas with no hair ECG pad electrodes have been quite useful. As most of them are made of silver, coated with silver chloride and are pregelled, they are quite handy for short-time recordings (Fig. 3).

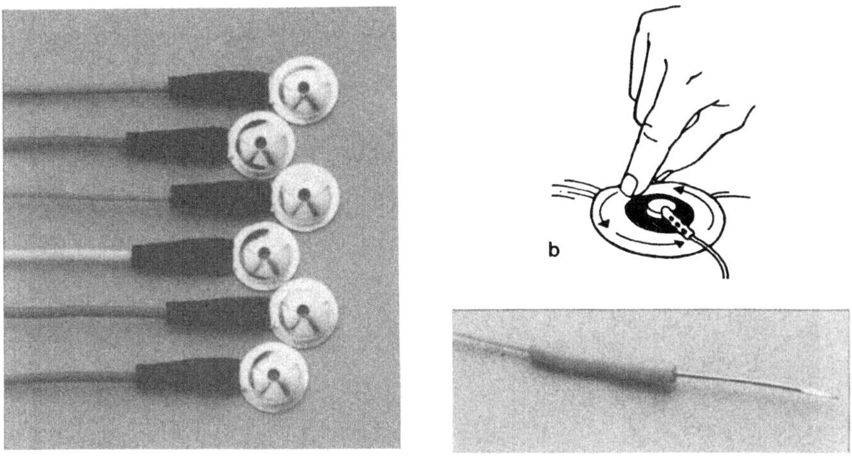

a c

Fig. 3. Examples of the various types of electrodes as they are commonly used for EEG recording: (a) cup electrodes, (b) pad electrodes, and (c) subdermal needle electrodes.

The purpose of the electrodes is to provide an efficient interface between the scalp potential and the insulated amplifier. The key to *electrode efficiency* is to minimize the impedance to electrical current flow. Excessive electrode impedance leads both to a decreased EEG signal and a frequency-dependent distortion. Therefore, *meticulous care* must be taken when applying the electrodes as they present the *prime source* of artifacts, interference and bad recording. The two major sources of excessive impedance, which is measured in 'Ohm', are the stratum corneum, the upper layer of the skin, and the skin-electrode interface. The upper layer of the skin, the stratum corneum, consists mainly of horn cells which act as a high resistance. This resistance can easily be reduced by mildly abrading the skin with an abrasive gel (e.g. Omniprep^{TM}). The skin-to-electrode contact can be widely increased by using a cup electrode with sufficient area of contact and applying enough conductive paste between skin and metal. Thus, normal electrode impedance should read below 5000 Ohm or 5 kOhm. Each electrode is filled with conductive electrode gel and attached to the scalp by a layer of collodion applied around the rim. Collodion provides a more secure seal than conductive paste and therefore is more satisfactory for intraoperative use.

The following specifications have to be met in order to facilitate recording in the OR or the ICU:

1. Dependable low impedance recording (below 5 kOhm) with a small disposable electrode to maximize signal input providing accurate display of cerebral cortical activity.
2. Simple and fast technique for electrode application reducing the need for specialized technical skill which will result in increased acceptability by those clinicians involved in cerebral monitoring in the OR and the ICU.

3. Specified areas of placement in order to compare EEG activities of the past with those of a present recording. The latter objective is further outlined in the following paragraph.

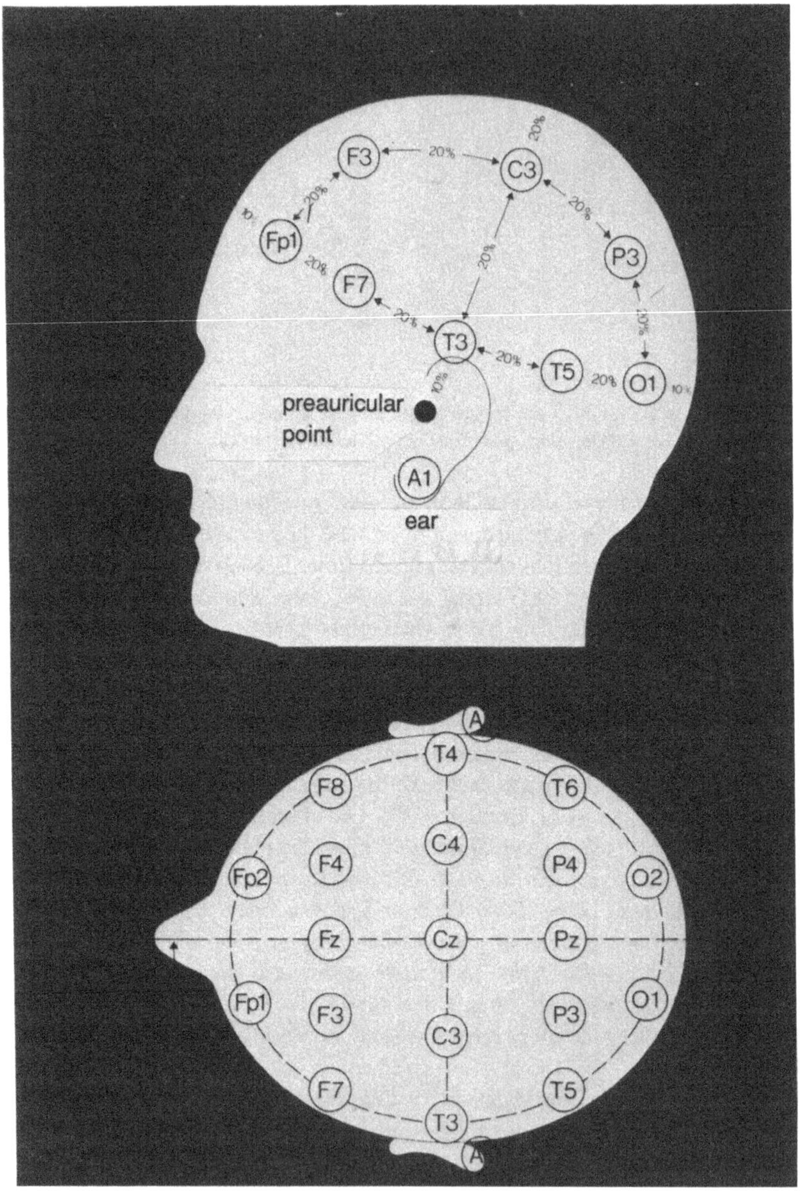

Fig. 4. The international 10/20 system for the placement of electrodes on the scalp in regard to different anatomical landmarks of the skull. (Photo: Siemens, F.R.G.)

2.2 Electrode placement

Standard neurological practice requires the precise placement of electrodes in relation to anatomical landmarks of the skull. The internationally accepted system for electrode placement is the 10/20-system (Fig. 4).

The International 10/20-System for standardized placement of electrodes has been developed by the 'International Federation of Societies for Electroencephalography and Clinical Neurophysiology' to permit comparison of recordings obtained in different patients with differently sized or shaped skulls [12]. The essence of the system is that the antero-posterior measurements are based on a distance between the *nasion* and the *inion* over the vertex of the middle bone, and fixed points, i.e. the frontal pole (Fp), the central (C), the parietal (P), or the occipital (O) point, which are marked along this line at percentages of the 10–20 range of the nasion-inion distance (Fig. 4).

Lateral measurements are based on the central, coronal plane from one preauricular point to the other. From here, at the 10–20% range, landmarks such as the temporal (T) point are set, where *even* numbers are used as subscript for points over the right hemisphere and odd numbers for those over the left hemisphere. Midline electrodes have the subscript z which stands for 'zero' (e.g.: F_z, C_z, P_z; Fig. 4).

2.3 Technical requirements to obtain a reliable EEG signal

Amplifiers. The EEG which is detected on the scalp is of fairly low voltage and is almost overflown by other physiological potentials such as the ECG or the EMG (electromyogram), or extraneous interference or noise of 50–60 Hz. To minimize these extraneous potentials as they are detected by the EEG electrodes, a *differential amplifier* is used. It has two inputs which represent the potentials at opposite polarities. This effectively cancels out both voltage signal common to both inputs, the common mode rejection, and amplifies only the voltage difference between the inputs (Fig. 5).

Electrical interference can be a major problem with intraoperative EEG monitoring since electrical signals from the brain are small and the operation room as well as the intensive care ward with their electrical power devices provide an 'electrically hostile' environment for recording. Thus, the EEG cables must be shielded and kept away from other electrical cords. The following points have to be kept in mind in order to obtain artifact-free EEG recording in the OR or the ICU.

Electrode leads are conductors. Think of them as antennas.

- Keep leads as short as possible.
- All electrodes should follow the same path to the preamplifier.
- Make sure that the protective insulation is not broken at any point.
- All electrodes should be of the same length.

To optimize electrical noise rejection in the recording room

- strive for an impedance between electrode and skin as low as possible,
- have the same impedance throughout the entire electrode montage,

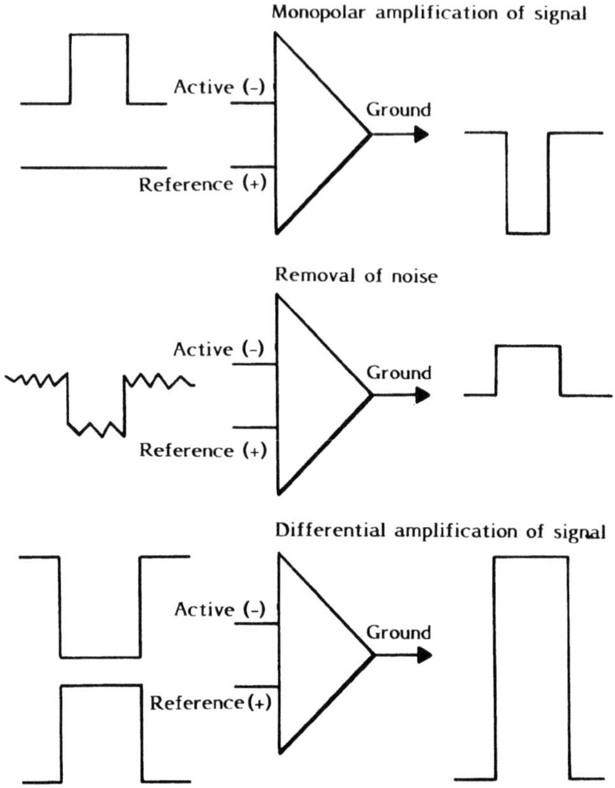

Fig. 5. The amplifier, due to its configuration, can do the following operations:
(a) amplification of the difference between active pole and reference,
(b) rejection of similarities between active pole and reference (common mode rejection, typically 100 db),
(c) signal points down at display when positive activity is at the active pole, and
(d) signal points up at the display when positive activity is at the reference pole.

- evaluate the choice of montage of electrodes (if possible no crossing of muscles), and
- select an appropriate amplifier gain for the signal.

Electrocautery still remains a problem, and recording must be interrupted during its use.

Filters. A technique to reduce noise content of the EEG takes advantage of the fact that the frequency content of the noise generally is different from that of the EEG. The use of frequency dependent filters balances judiciously the reduction of noise against decreased fidelity of true waveforms. In Fig. 6 noise is gray containing a random mixture of all fequencies, whereas EEG waves contain only selected frequencies. Thus, the objective of filtering is to remove the noise which does not overlap the signal by preventing it from entering the computer system. The sharp edges as shown in the figure below represent the lowcut and the highcut site where noise frequencies are blocked. The actual filter system, however, has a gradual slope so that noise is *not* totally eliminated.

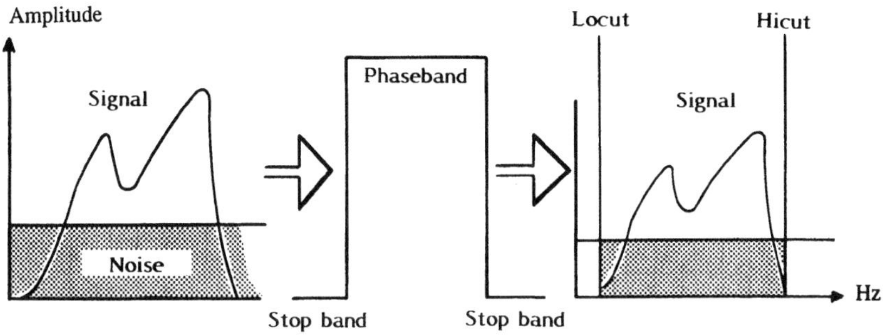

Fig. 6. The principle of a filter to remove electrical noise from relevant EEG frequencies.

For standard clinical EEG reading in the OR and the ICU filters usually are set to attenuate signals slower than 0.5 Hz and faster than 30 Hz. This band pass represents the best compromise for EEG reading in an environment which is characterized by a concomitant increase of electrical noise.

2.4 The recording procedure

A combination of electrodes used for cerebral monitoring is referred to as a montage. The monitoring chosen for a given situation depends on *what area of the cortex* is to be examined and *what information is being sought.* Basically, there are 3 types of montages:

1. In the *bipolar montage* a sequential connection of channels along the antero-posterior or transverse rows is used. Thus the potential difference between the two electrodes is measured.
2. In the *monopolar* or common reference montage each channel records between active scalp electrode and an 'indifferent' electrode usually on the chin, forehead or the ear.
3. Joining all scalp electrodes to a common resistor of equal value, i.e. *the average reference*, and measuring between this common reference and each channel, the *average reference* monitoring is used.

Commonly in anesthesia and in critical care *one bipolar* montage of each hemisphere is used as this montage is able to:

- measure the difference between both hemispheres,

– evaluate global ischemia of one hemisphere, and
– apply a limited number of electrodes as fast as possible.

Additionally, in cases of ischemia due to carotid occlusion, widespread frontal and occipital EEG changes can be observed. This suggests that gross events can be monitored *without* complex montage [25]. Also, if electrodes are placed too close to each other, they do not necessarily monitor the activity directly beneath or between them [26]. For these reasons the montage F_3-mastoid, behind the left ear, and

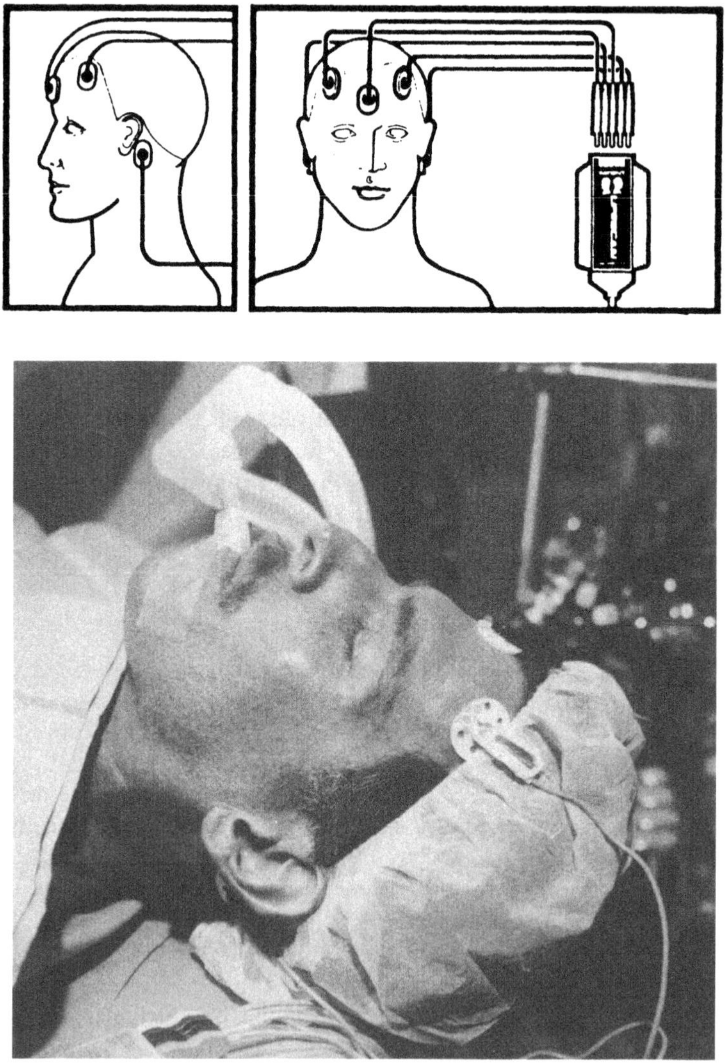

Fig. 7. Conventional ECG stick-on electrodes are placed on the forehead and mastoid process, simplifying patient preparation. For a symmetrical picture, electrode location is as follows: One on each frontal area (F_3 and F_4 respectively). One behind each ear lobe (A_1 and A_2; the ipsilateral mastoid) and one on the center of the forehead (FpZ).

F_4-mastoid, behind the right ear, taking FpZ as the reference point, is sufficient to evaluate gross changes in cortical activity of both hemispheres which are either due to the anesthetic or due to hypoxemia (Fig. 7; see also Fig. 4).

2.5 Display techniques of the EEG

Strip chart is the conventional technique used in clinical neurology. A strip chart is written by an ink pen and moved by a paper drive. At a proper speed of 30 mm/sec the machine produces 108 meters/hour of recording paper! It is for this reason that computerized techniques have been developed to *process and compress* EEG data in cerebral monitoring. There are many methods for processing the EEG, but the most widely used method in intraoperative monitoring is *frequency analysis*. In this technique the EEG is seen as a simulation of simple sine waves with different frequencies. Frequency analysis mathematically decomposes this complex of sine waves with different frequencies into the various components using *Fourier analysis*. It was not until the rediscovery of the Fast Fourier algorithm by Cooley & Turkey in 1965 [9] that it became possible for computers to perform *on-line* power spectral analysis of the EEG. This mathematical technique is the fastest because of the fewer calculations in carrying out frequency analysis. Thus, a signal from 2–60 sec, depending on computer capacity, is taken and frequency analysis is performed (Fig. 8) using *Fast Fourier Transform (FFT)*. The resultant analysis is recorded graphically; the power of the signal at each frequency is plotted vertically against the frequencies from 0–30 Hz. This produces a power spectrum of the original EEG signal.

In the FFT mode the computer samples the EEG signal *periodically* over a certain period called epoch and converts that sample to a digital number of finite precision. The FFT method uses a fixed number of samples in that signal. At a 2 sec epoch, for instance, 128 data points are sampled per second, thus yielding 256

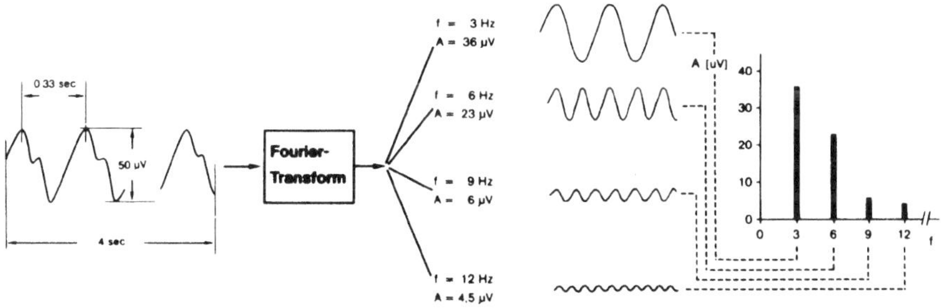

Fig. 8. Principle of Fast Fourier Transform (FFT) of the raw EEG. The original EEG wave is put into a digital format by sampling it repeatedly at very small intervals. The data of this epoch is then passed to the computer to perform FFT. As a result the amplitude of each frequency band is derived. In a final step the amplitudes are squared to give power, and plotted on a histogram.
f (frequency) = 1 Hz; A (amplitude) = 50 μV.

data points per epoch. Thereafter the distribution of energy or power at each interval of frequency is computed which results in the frequency spectrum of that sample (Fig. 8).

A procedure as outlined above produces a power spectrum of the original EEG signal. The approach has been further expanded by plotting successive power spectra vertically so that changes over longer periods of time can be followed. The *Compressed Spectral Array (CSA)* format [20] combines high data compression and compact trending at the base activity and of the distribution of dominant frequencies and their evolution in time (Fig. 9).

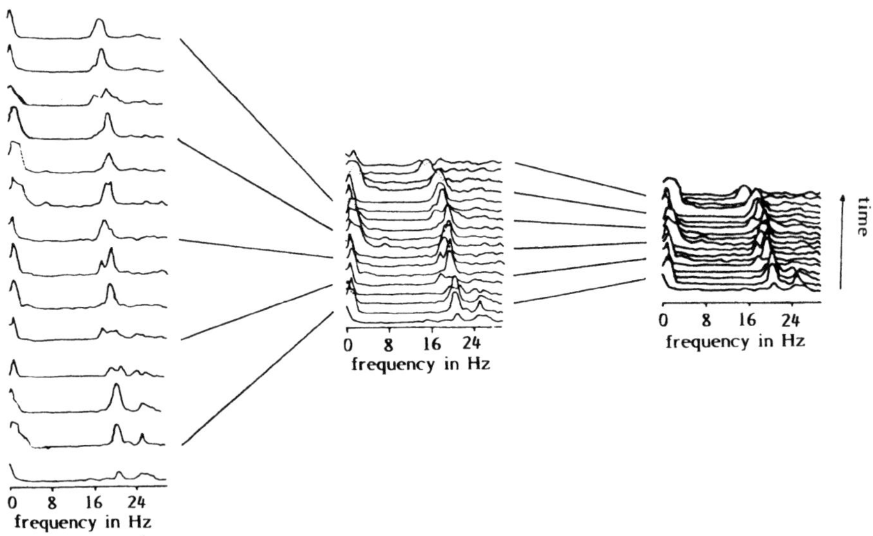

Fig. 9. Individual epochs of analysis are plotted closely above each other resulting in a 'hill and valley' pattern. Such a picture is called 'Compressed Spectral Array (CSA).

In the CSA mode sequentially calculated spectra are drawn with the aid of a plotter in a stack, each in front of the next. A hidden line suppression algorithm is usually implemented in the display to remove spectral information behind background feature to add a three-dimensional effect.

A further simplification of EEG interpretation may be made by having the computer quantify certain aspects of the computer spectra:

- Peak Power Frequency: the frequency at which the highest power is found to occur in the current spectrum (Fig. 10).
- Median Power Frequency: the frequency in the current spectrum at which half the power is above and half is below ([21]; Fig. 10).
- Spectral Edge Frequency: the highest significant frequency present in the current EEG spectrum. This usually is the frequency which represents 95% of all the power ([22]; Fig. 10).
- Density Modulated Spectral Array (DSA [23]). In this technique the vertical axis is frequency and the horizontal axis is time. The relative power in the various

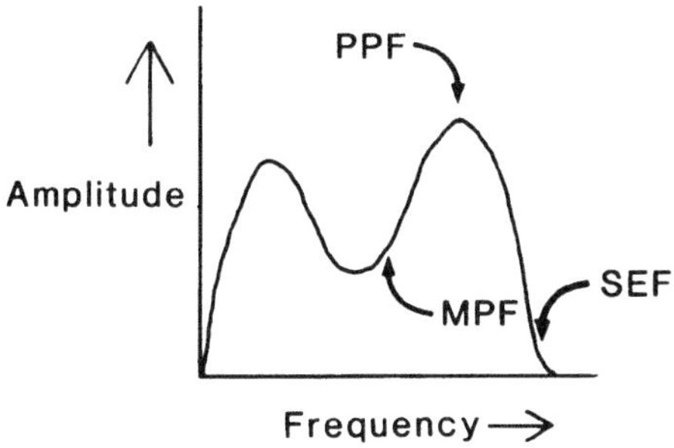

Fig. 10. The various attempts to simplify analysis of EEG power spectra.
Median Power Frequency (MPF) is the frequency which bisects the power curve. Peak Power Frequency (PPF) is the frequency which comprises the largest single component of the power spectrum. Spectral Edge Frequency (SEF) is the highest frequency component in the EEG.

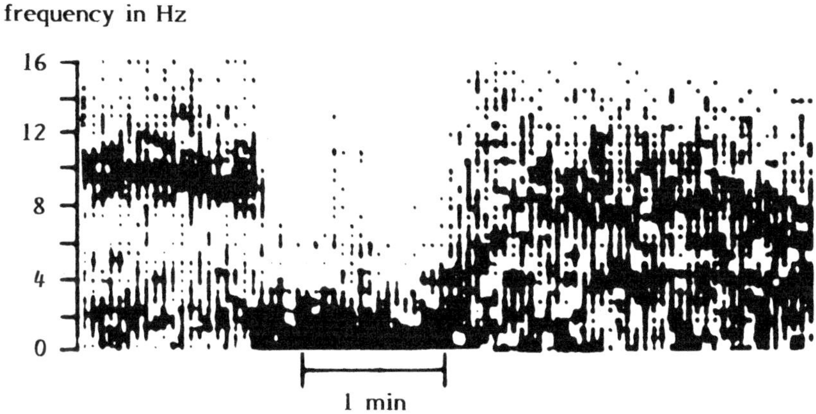

Fig. 11. Density Modulated Spectral Array (DSA) as it appears on the plotter.

frequency bands is reflected in the density of dots. As time passes, a dot pattern develops with changing power content (Fig. 11).

So far all the univariate descriptors for intraoperative representation of cortical waves have been shown only to be accurate when the EEG behaves as an unimodal distribution of frequencies changing slowly with time. However, in the majority of cases these univariate descriptors of the EEG appeared inadequate to demonstrate changes of the EEG during anesthesia [14]. Further development in displaying the processed EEG in a three-dimensional manner with color delineation of the various frequency bands improved and aided EEG interpretation, and was realized in the Lifescan^{TM} EEG monitoring device (for further details see page 71).

2.6 Artifact rejection during processed EEG measurement

An artifact is a sudden excess in voltage with respect to the average signals already occurred which, however, may not exceed the units input dynamics. Artifacts over 30 Hz such as

- power outlet noise, and most of all
- muscular potentials from the tissue underlying the scalp (i.e. fronto-temporalis montage),
- unsteady electrodes, and
- rapid occular movements

are detected by an artifact rejection unit and identified as such.

The *electrosurgical* or *bowie noise* is one of the most common generators of electrical noise. Bowie noise appears in the EEG as a complex mixture of very high speed spikes and periods where the input amplifiers are saturated producing a flat line at one end of the scale. The spectrum may resemble that of motion artifacts, or it may appear in a diffuse high power distribution across the entire spectrum. Interferences coincide with the application of the cutting or coagulation current.

The following problems may be encountered during EEG recording in the OR, all of which interfere with the original EEG wave and its analysis with the aid of a computer:

1. Electrical interference:
   a) improperly applied electrodes,
   b) electrocautery,
   c) cell-saver machine, and
   d) line interference;
2. Equipment failure:
   a) major malfunction of the processor, and
   b) malfunction of accessories.

2.7 Signal conduction and conversion

Most of the processed EEG machines used nowadays run in a similar manner with three phases in their operation cycle:

1. EEG conduction and conversion,
2. Data analysis, and
3. Formatting and display.

The EEG signal is collected from the scalp by disk, needle, or ECG electrodes applied in the same manner as for routine ECG examination. The low level EEG signal which runs between 10–100 μV is amplified, filtered and checked for artifact and sent to an analog to digital converter.

The amplifiers are low-noise, high-gain differential amplifiers with a high impedance front end. After amplification the signal then is filtered by a third order low-pass filter with a cutoff (2 db down) at 30 Hz and by a fast-order high-gain filter with a cutoff at 0.5 Hz (high pass filter) giving a passband of 0.5–30 Hz. The filtering serves to reduce the noise picked up along with the EEG by the sensitive amplifiers. Two EEG channels then usually are multiplexed and sent to the analog-

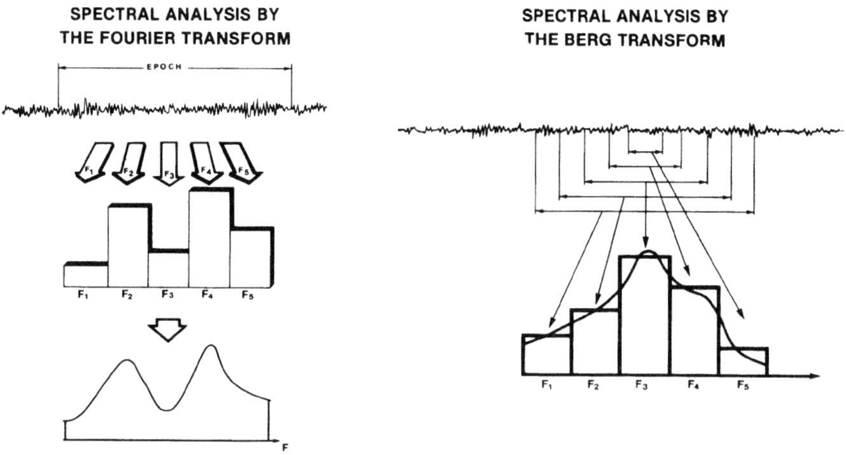

Fig. 12. The difference between FFT and Berg Transform of the EEG signal resulting in power spectra.

to-digital converter on a first procedure board. The incoming signal is sampled at 256 Hz giving a sampling rate of 128 Hz per channel. Samples are converted to bit words, separated into two channels and stored by the microprocessor. Artifact detection is performed by means of a three-part algorithm which checks for:

1. EEG signals too high for the amplifiers.
2. A zero derivative signal, and
3. 60 or 50 Hz noise.

The presence of artifact is flagged in several ways on the screen in each display mode for each channel. The raw EEG signal is available for display on the screen whenever a specially designated key on the keyboard is pressed. Most devices analyze the raw EEG using the Fast Fourier Transform (see Fig. 12), while only one device uses the Aperiodic Analysis. The resulting spectra along with the raw data are sent to the display processor.

In the Fast Fourier Transform (FFT) mode, the spectral results of epochs 2 second each (=256 data points) are used. Contrary to the FFT episodes, Aperiodic Analysis of the EEG [26] yields an estimate, where amplitude and frequency are presented on a wave-by-wave basis (Fig. 13). By the latter method, a wave is defined as a fluctuation in voltage that occurs between two local minima in voltage. Thus, the algorithm scans through time for decreasing voltages. When an increase in voltage is noted relative to the previous point, then the previous voltage is noted to present a minimum valley V_1 while scanning along as the voltage increases.

For any increase in voltage thereafter relative to the previous time point, the voltage value represents a local maximum peak P (Fig. 13). To detect slow delta or theta waves in the presence of high frequency waves, a slow wave method is used similar to the technique as described above [27], except that it scans for the largest valleys and peaks between *zero voltage crossing* (Fig. 13).

As each channel samples the signal 960 times per second compared to the usual

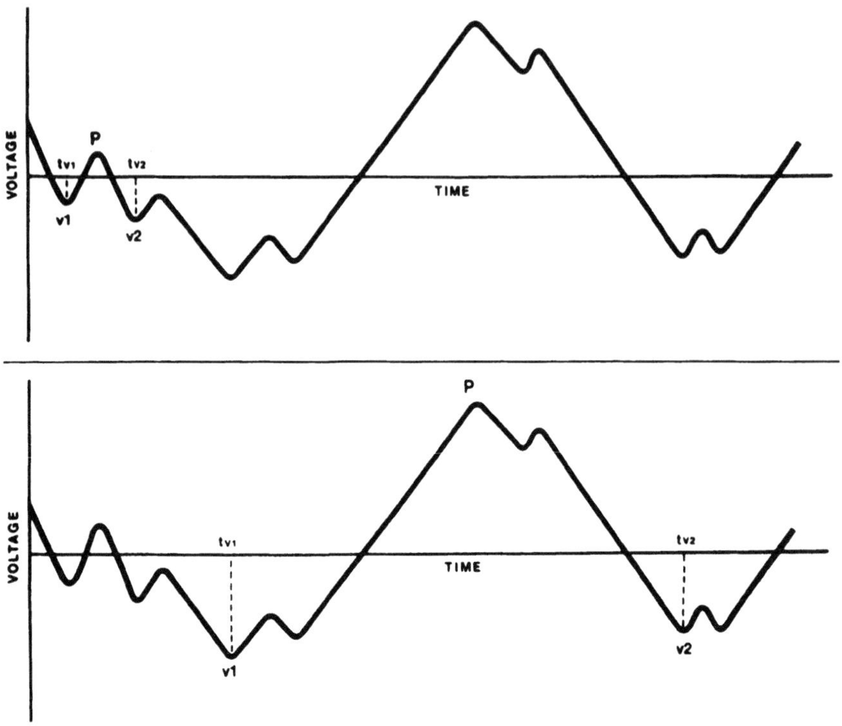

Fig. 13. Principle of the algorithm of Aperiodic Analysis of the EEG. The amplitude of the waves is defined as the average difference in voltage between P and V_1 and the difference in voltage between P and V_2. The power as it is visualized in a vector $(\mu V) = (VP - V_{V_1}) + (V_P - V_{V_2})/2$. Frequency of the wave is defined as the inverse of difference in time between minimal V_1 and V_2. Thus frequency $(Hz) = 1/(t_{V_2} - t_{V_1})$. P = peak; V = valley.

128 used by most of the algorithms, Aperiodic Analysis seems to result in better representation of EEG waves of short duration. Thus, fast waves with little energy which, however, may be important in the clinical setting, are preferred. By comparison, FFT obtains a spectral representation of the EEG, which, due to the stochastic nature of the EEG waves, is not too logical. As EEG waves do not seem to behave like a mixture of regular sinoid-waves, but more like random non-periodic fluctuations, other forms of computation of power in the EEG are likely to be more rational. Additionally, the finite time of epochs during which data is collected with the FFT reflects spectral representation, missing those periods at which a change may have happened. A technique with an averaging, smoothing mode, alters the accuracy and resolution of the resultant frequency spectrum [28].

All EEG data can be formatted and displayed in one of the 3 different formats when using the FFT or in one of the 2 different formats when using the *Aperiodic Analysis*:

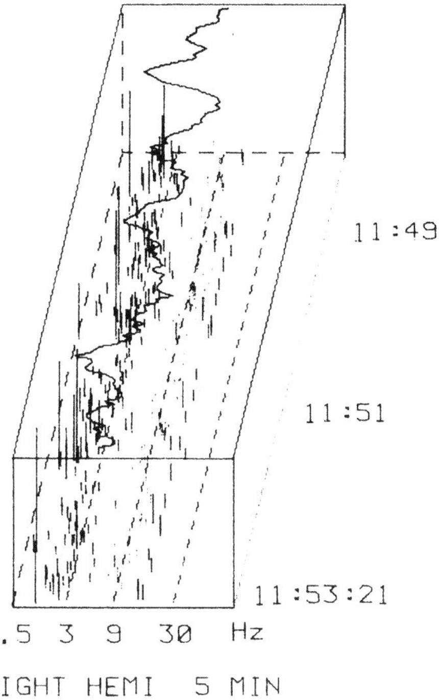

Fig. 14. Representative example of analyzed EEG waves taken postoperatively from a patient after enflurane anesthesia and induction with midazolam. Aperiodic Analysis denotes fluctuation of vigilance which is best seen by the spectral edge as it changes along with time.

1. The raw EEG signal.
2. A bar graph of each channel power spectrum (FFT algorithm), or an amplitude and frequency information as used in the two-dimensional screen with a color vector *and* a three-dimensional time axis into the depth of the screen as time moves on (Aperiodic Analysis; Fig. 14).
3. Compressed Spectral Array (CSA) with power spectral lines taken at successive time intervals on the screen to produce a quasi three-dimensional trend (Fig. 15).

The display mode of both methods has the following number of options:

1. Display of a single channel or both channels of EEG waves on screen.
2. Updating the time from a two-second analysis to several minutes.
3. Storage of analyzed data up to several hours.
4. Display of Spectral Edge Frequency (SEF), the location in the power chart where the processor draws a line reflecting a point (Hz) in the spectral content which incorporates a certain amount (usually 95%) of total power.
5. Change in the gain setting for the power values in the spectral histograms.
6. Menu-driven check of impedance of all electrodes pasted to the scalp.
7. Display of power bands and/or power vectors in the various frequency bands.
8. Computation of the absolute power values (nanoWatt, microVolt $\times$ 2) which can be printed by a built in plotter.

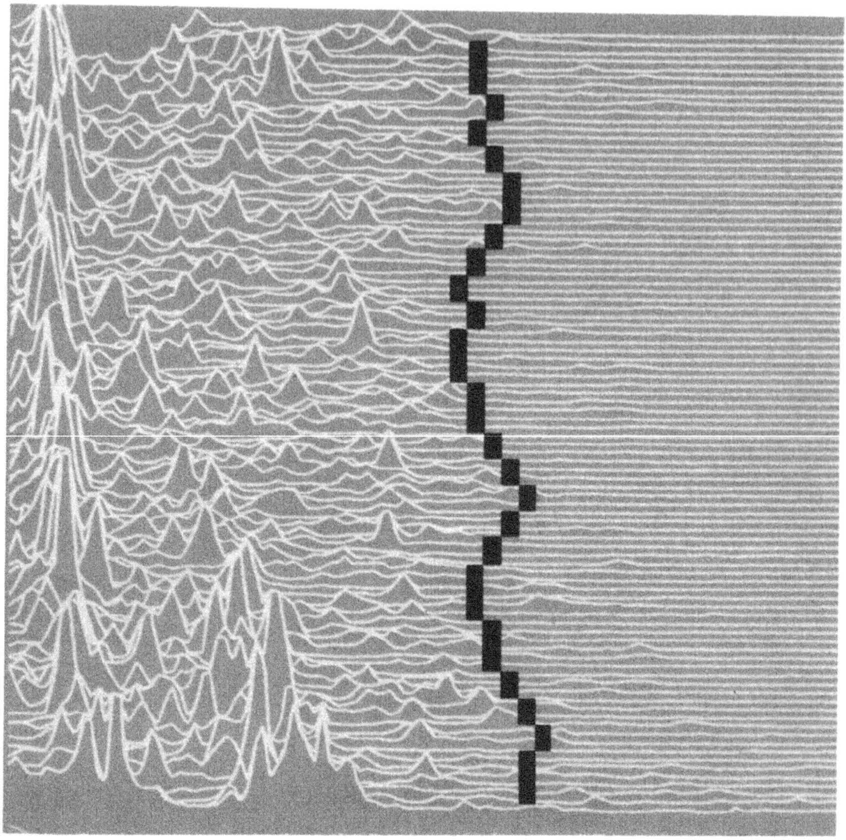

Fig. 15. **Representative example of EEG waves analysed by FFT and formatted in CSA. The Spectral Edge Frequency (SEF) is denoted by the darker bars on the right-hand side.**

9. Extension of cerebral monitoring to the *Evoked Potential Mode* in order to evaluate the function of somatosensory pathways in deeper (subcortical) brain structures.
10. Use of an EEG analysis device as a system to process raw EEG signals from tape or a regular EEG machine.

Basically, computerized EEG recording consists of five subunits:

1. *The preamplifier* which is connected to the processor via lines and enhances the brain waves to take up the electrodes of the patient.
2. *The microprocessor* for data conversion, analysis, and formating.
3. *The display board or screen* containing a screen for black and white or color processing of results.
4. *The printer or plotter* for permanent plotting of relevant data.
5. *The keyboard* with numbers and/or different color coded digits to operate the monitor and select the montage.

3. Set-up of monitoring the EEG

The electrode montage

This chapter describes the cable connectors and switch of some of the most often used EEG monitors in the OR and the intensive care.

Before one sets about monitoring the brain, it is important to note that the 12 to 16 lead EEG usually used in neurology and psychiatry, is not mandatory during anesthesia and critical care. Since localized changes are of no significance to the anesthesiologist and/or the intensive care therapist, 'only' 2 leads, to the right and the left hemisphere, are necessary to obtain a picture about global changes that occur in the brain. Thus, the effect of anesthetics in the brain and of ischemic events can be recorded with sufficient ease for routine use. The computer-run monitors have to fulfill basic requirements in order to become a useful tool for the clinician. It should:

1. be compact to fit into the OR and the intensive care wards, designated to share the sparse room with other electronical equipment,
2. be mobile on rollers so that it can be easily wheeled from one place to the other,
3. be simple in operation, preferably using touch keys,
4. contain a view screen with sufficient resolution in contrast for changes to be viewed from a distance,
5. contain an input/output terminal for system expansion (preferably an RS-232 serial port),
6. contain memory capabilities for at least 12–24 hour recording of patients in the intensive care,
7. allow storage of snapshots of EEG events for later retrieval via the built in plotter or printer,
8. monitor the impedance continuously in order to affirm sufficient electrode contact,
9. contain on-screen a self explanatory user guide after switch on,
10. store, for later retrieval and comparison, a number of typical preprocessed events that are stored in the data bank permanently, to aid the clinician in decision making, and
11. have the option of permanent data storage on casette or floppy disk.

In order to facilitate and speed up electrode positioning, three frontal EEG electrodes at positions F_3 (+) and F_4 (+) with the reference at position F_pZ (forehead) are used while the negative pole (–) is applied on the mastoid of the respective side (Fig. 16).

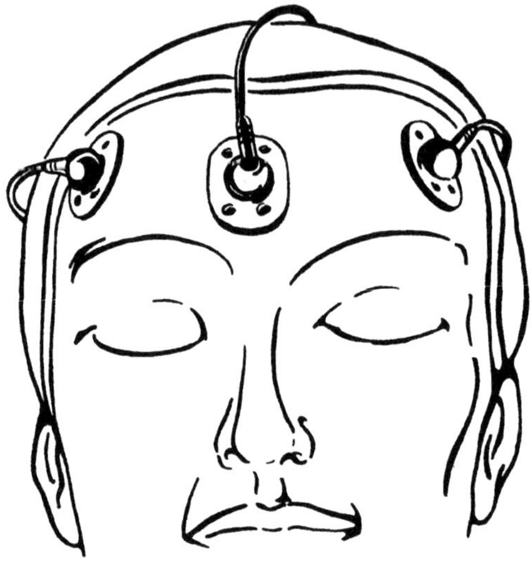

Fig. 16. Electrode montage on the forehead for recording of EEG activity.

Care should be taken in the preparation of the skin underlying the electrodes:

a) select hair-free skin site,
b) clean site of application with a mild abrasive cream (e.g. Omniprep™) using a cotton tip applicator in circular motion to abrade the skin,
c) use alcohol to wipe off the abrasive cream and to further clean the site of application,
d) next use a dry gauze and rub briskly to dry the area,
e) ensure that the skin is *completely dry* before applying the self-adhesive EEG electrodes,
f) remove the release band of the electrode taking care to minimize finger contact with adhesive areas,
g) ensure that the center pad is sufficiently moistened with electrode gel, and
h) apply the EEG electrode utilizing finger pressure around the adhesive area in a circular motion (Fig. 17),
i) snap the shielded leads to the electrodes,
j) connect the colored leads with the similarly colorcoded sites on the preamplifier,
k) check impedance [If unacceptable impedance occurs, identify the electrode which might be responsible and replace it by repeating steps (a) to (k); if impedance is still high, one lead must be faulty and needs to be replaced.],

l) clamp the preamplifier to the bed linen *near the patient*,
m) check for crossing of shielded leads with other electrode wire (e.g. ECG) to avoid interference, and
n) use the touch key for activation of the *raw EEG signal display*. Check if both channels show a good reading without the EEG waves wandering over the screen.

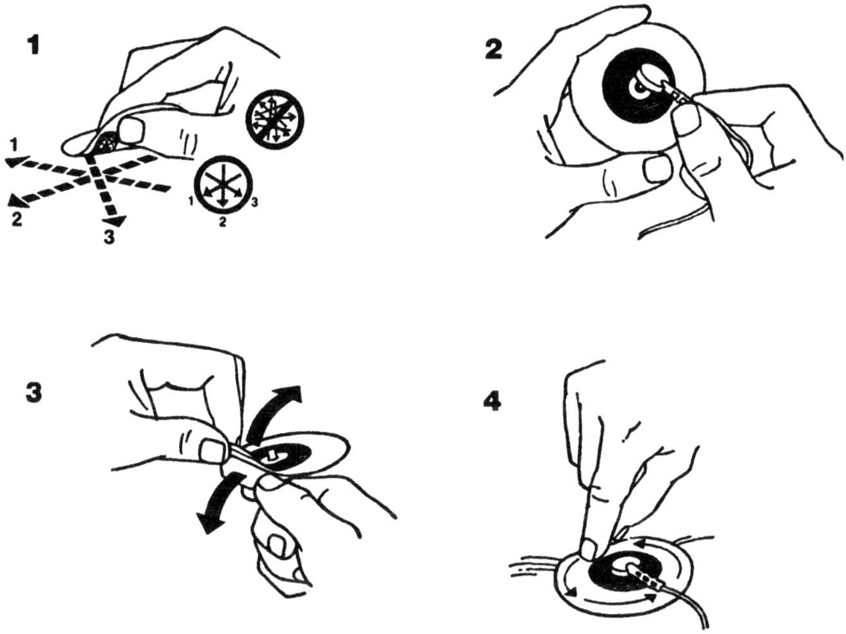

Fig. 17. Procedure as how to accurately apply the pad electrode to the skin.

Only now can the observation of the processing mode of the EEG begin. For optimal evaluation of cortical activity and to guarantee an artifact-free recording over a long period, every time the EEG is monitored and the raw signal is processed for data comprehension and reduction, steps (a) to (n) must be followed *meticulously*.

It is important to point out that the EEG '*only*' reflects potential changes between the two selected electrode sites of upper cortical layers. The EEG *does not* reflect functional deficits or changes in the deeper, subcortical structures. (For the evaluation of deeper brain structures see Chapter 9.)

Remember: dry skin =
monitoring of positive skin contact =
low impedance =
reliable signal!

Thus, in cases where the patient is perspiring profusely, self-adhesive electrodes tend to lift off, resulting in a deterioration of the raw EEG signal and a wandering base line, which in a number of cases is detected by the algorithm as slow EEG activity and *not* as an artifact (!). In such circumstances it is advisable to use silver/silver chloride cup-electrodes which are fixed to the skin by means of a special glue (collodion). (See application of cup electrodes in Chapter 10 on Evoked Potential monitoring.)

4. Anesthesia and the EEG

4.1 Rationale for the use of the EEG in anesthesia

The use of the EEG in anesthesia originally evolved from the hope that it would be possible to define the anesthetic depth by reference to the EEG pattern, irrespective of the agent being used. Of greater importance, aside from the ease of interpreting the EEG signals with the above-mentioned menu-driven computations is, however, the speed of response to ischemic events which can be visualized on the EEG-screen. Such events can occur very rapidly during surgery, at a state where anesthesia *covers* the events to such a degree that the sequelae will only become aware when the patient regains his consciousness. The common EEG-pattern sequences as they occur with increasing depth of anesthesia are as follows:

1. In the resting state the marked variation that exists in the EEG among awake, normal subjects disappears, and
2. the EEG tends towards a common pattern in all anesthetized patients. Additionally, it can be noted that
3. the changes under anesthesia are generalized, so that electrode placement on the scalp is *not* critical and in general a pair of widely spaced electrodes will be sufficient to reflect a representative record. All these facts greatly simplify EEG comparison under anesthesia, Also,
4. anesthetic agents initially increase electric output of the brain and, after a peak is reached, there is progressive reduction with increasing anesthetic depth (early burst suppression), until finally, the EEG becomes totally flat, devoid of any activity (late burst suppression) to total EEG suppression (Fig. 18).

However, some volatile anesthetics, especially those of the ether type, develop a different pattern beyond suppression which is characterized by generalized seizure activity. In modern anesthetic techniques using the type of balanced anesthesia with different compounds, interpretation of the EEG in regard to the anesthetic depth tends to become difficult. This is primarily due to the combination of drugs affecting different levels of the brain. Thus, absolute levels of anesthesia are difficult

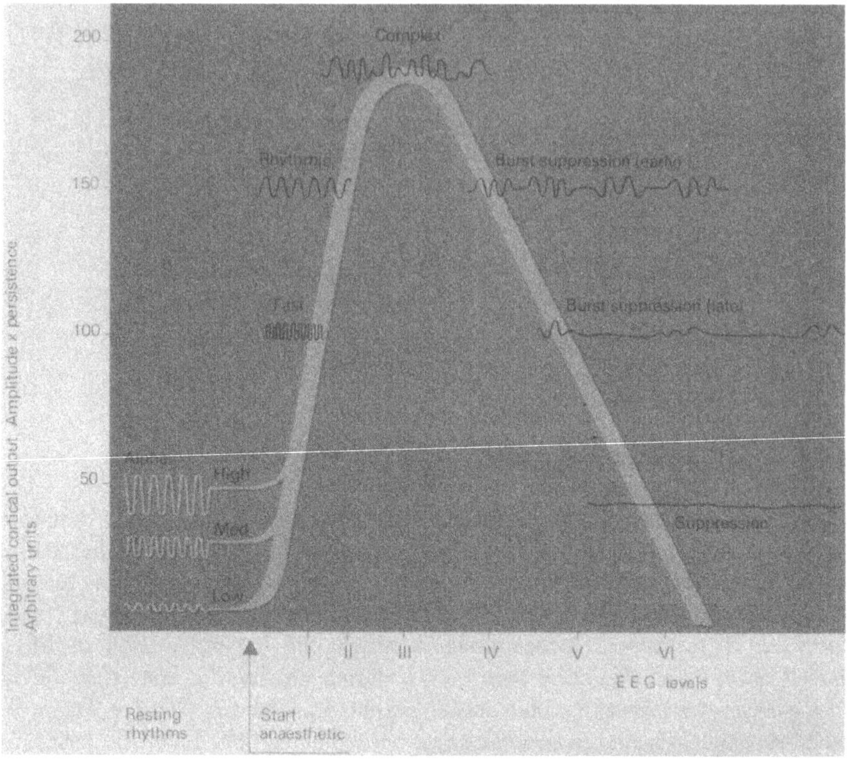

Fig. 18. EEG pattern as seen with increasing depth of anesthesia. The graph shows the change in integrated voltage output related to pattern changes. With certain ether agents such as enflurane, a stage of seizure activity may follow the suppression (adapted from Stockard & Bickford [29]).

to assess with the use of multiple agents on board, however, shifts of the depth of anesthesia can almost always be seen.

4.2 Inhalational agents and their effect on the EEG

N_2O. Up to 30% of nitrous oxide in the inspired air results in minimal changes in the EEG in spite of the marked psychological disturbances that can be observed. At 50% unconsciousness develops with increasing disappearance of alpha rhythm with interposed short episodes of fast beta activity. This is followed by the appearance of slow theta (4–7 Hz) activity, while at 80% N_2O fast activity disappears totally and only theta remains in the EEG picture [30–32].

Halothane. Subanesthetic levels of halothane produce a fast 12–18 Hz pattern in the EEG which continues as consciousness is lost. At 1 MAC (=0.84 Vol%) the

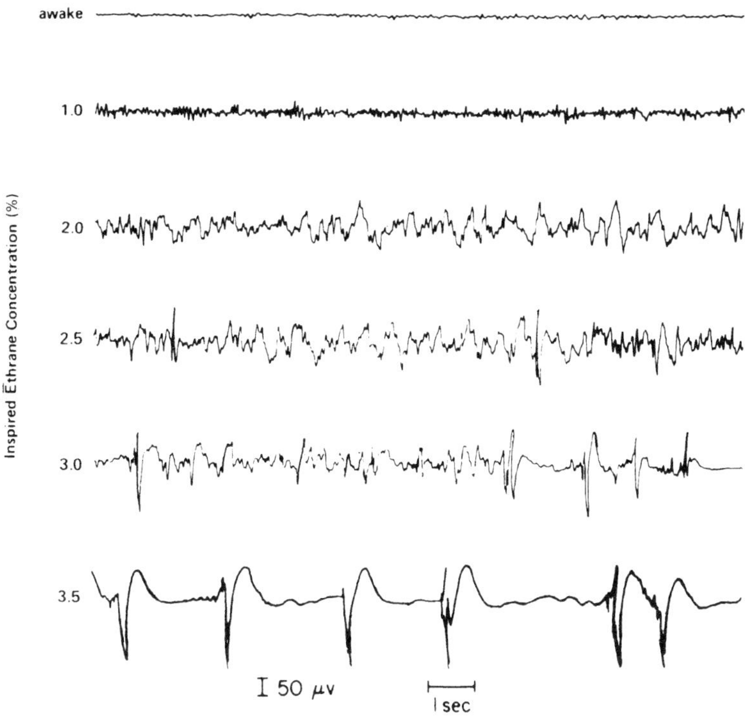

Fig. 19. Increase in inspired enflurane concentration results in a decrease of frequency and increase of amplitude of the EEG. At 3–3.5 Vol% burst suppression occurs. At 3.5 Vol% spike and dome activity may be observed (adapted from Neigh et al. [35]).

dominant activity falls between 11–16 Hz. As anesthesia deepens there is a near-linear decrease of EEG rhythm by about 1–1.5 Hz with each 0.5 MAC (=0.42 Vol%) halothane concentration. Above 2 MAC, EEG power in the delta band arises from background activity until finally at 4 MAC, almost all activity is centered around the 0.5 Hz level [29]. If nitrous oxide is given together with halothane, the degree of slowing in the EEG is augmented. Thus, unconsciousness occurs when the EEG activity is centered around 12–15 Hz (=alpha), while surgical stimulation is tolerated at peak power frequencies between 6–8 Hz (=theta) [33, 34].

Enflurane. The electroencephalographic changes seen with enflurane are in many ways identical to the changes seen with the other inhaled anesthetics: an increase in the concentration decreases frequency and increases voltage [29]. At concentrations of 1 MAC (=1.7 Vol%) the EEG dominance is of slow waves with high amplitudes in the 4–8 Hz range. The waves further slow as anesthesia deepens while early burst suppression periods occur at 1.5 MAC. This is especifically the case when lowered pCO_2 levels are present; the EEG picture may be interrupted by the appearance of occasional high voltage *spike waves* (Fig. 19). In deep enflurane anesthesia (2–3 MAC), epileptic-like single bursts or groups of bursts become apparent which are followed by 5–15 seconds of isoelectricity, associated with or without abnormal muscular movements. The latter manifests itself by twitching of individual muscle groups [29, 35, 36].

Isoflurane. Pharmacologically, this volatile anesthetic agent is a structural isomer of enflurane, which induces a similar EEG pattern as enflurane (Fig. 20). However, this halogenated methyl-ether does *not* elicit the burst-like pattern in the EEG at 2–3 MAC (=2.4–3.6 Vol%). With deep isoflurane anesthesia there is dominant power in the delta range which may be followed by electrical silence of several seconds [36].

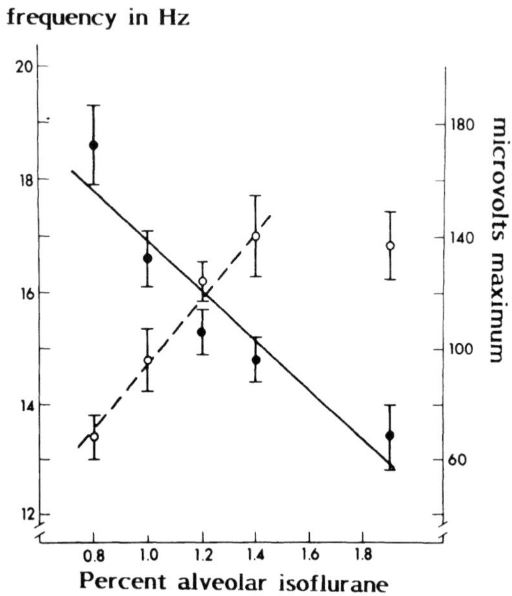

Fig. 20. Amplitude (open circle) and frequency (closed circle) changes in the EEG with increasing concentrations of isoflurane (adapted from Eger et al. [37]).

The capacity of isoflurane to produce an isoelectric electroencephalogram and the absence of seizure activity, sets this volatile compound apart from the other inhaled anesthetics. Although ether also can produce an isolectric pattern, the concentration required (in MAC multiples) is significantly greater. In addition, ether agents can produce a convulsive EEG pattern which may appear at concentrations only slightly higher to produce an isoelectric pattern (Table 1; adapted from Stockard et al. [29]).

The effect of loss of EEG activity during anesthesia can be considered as a sign of deep anesthesia. In a representative example of a white female 68 years of age undergoing gall bladder removal, enflurane-N_2O/O_2-anesthesia with 2 Vol%

Table 1. MAC multiples required to produce an isoelectric and seizure pattern in the EEG of man with different anesthetics.

	Isoelectricity	Seizure
Isoflurane	2.0	0
Ether	3.5	4.5
Halothane	3.5	0
Enflurane	0	2.0–3.0

produced a dominance of delta waves in the EEG power spectrum (Fig. 21). This is clearly marked by the Spectral Edge Frequency (SEF) which centers around 3 Hz. At 12.29 hours without any change in blood pressure and/or heart rate, EEG activity suddenly dropped. This is depicted in the power spectrum with a drop in the SEF to 0.5 Hz and a marked depression of amplitude in the delta range. Subsequently, enflurane concentration was reduced to 0 Vol%, which resulted in the

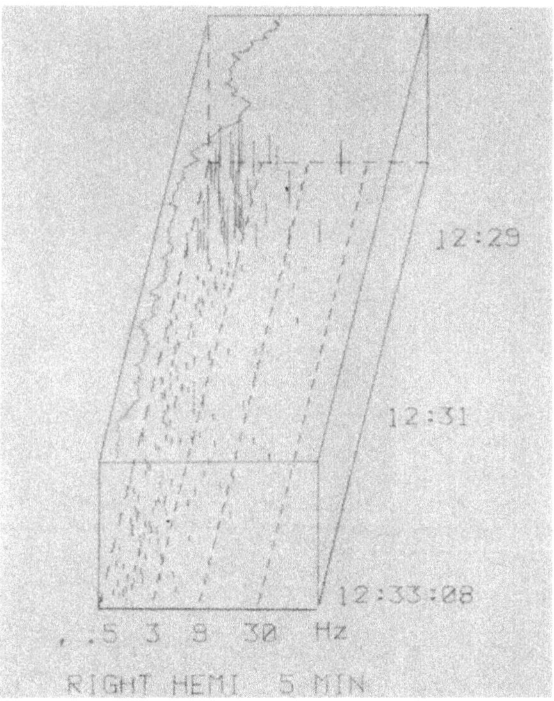

Fig. 21. Example of intraoperative EEG monitoring where a deep anesthetic plane was detected and corrected (EEG power spectra derived with the Lifescan^{TM} cerebral monitor).
x-axis = frequency band (Hz)
y-axis = amplitude height (gain 100 μV)
z-axis = evolution in time (minutes)
Ethrane 0 vol%; N_2O/O_2 = 2/1; Edge 0.5 Hz

reappearance of power in all frequency bands. At the end of the surgery the patient recovered uneventfully. Without the use of the EEG monitor the deep anesthetic plane would not have been discovered and eventually would have led to cardiovascular depression and/or prolonged awakening from anesthesia.

The difference of both structurally similar anesthetics, enflurane and isoflurane, on EEG activity can also be seen in patients undergoing CABG (coronary artery by-pass grafting) with *hypothermia.*

During the cooling period there is a temperature-related depression of the EEG, in spite of constant blood anesthetic concentration. During rewarming and continuous enflurane administration, the EEG power in the theta band (4–8 Hz) is characterized by a marked increase at a point where body temperature reaches 30 °C. This is in contrast to isoflurane anesthesia. Rewarming of body temperature is characterized by a gradual increase of power in the theta band. The former may be interpreted as 'benign convulsive activity' with a mean power difference of 345% in the isoflurane as compared to the enflurane group (adapted from Freye et al. [39]; Fig. 22).

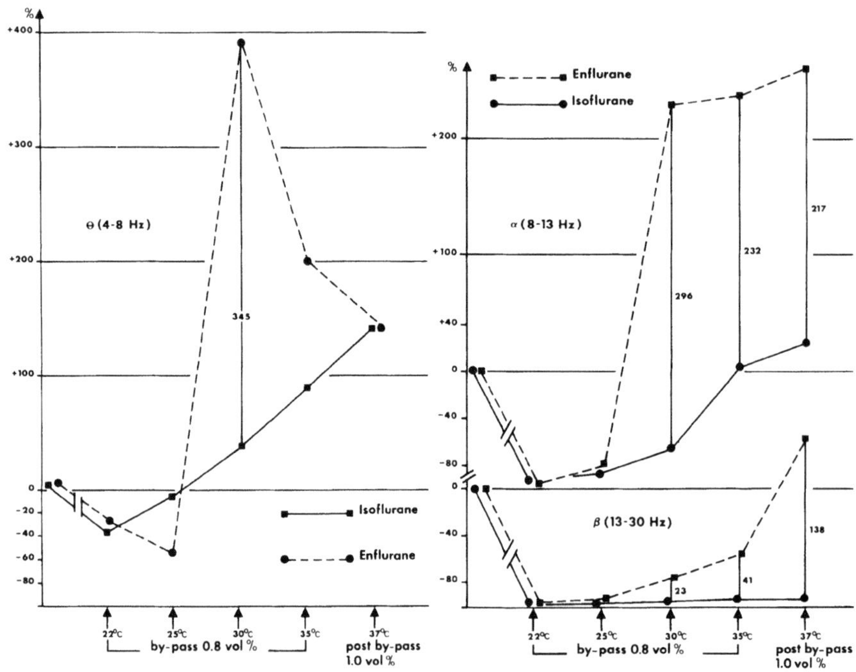

Fig. 22. Effect of rewarming on the various EEG power frequencies of the EEG in patients undergoing AC by-pass operation with hypothermia. Anesthetic blood levels were constant throughout the rewarming period. EEG power spectra were derived with the Neurotrec^{TM} Cerebral Monitor. Note the rapid increase of power in the theta- and alpha-band in patients (n=10) receiving enflurane- compared to patients (n=10) being anesthetised with isoflurane-oxygen anesthesia.

The increase of power in the theta band during rewarming is far greater than the increase of power in the faster alpha and theta domain, the cause of which is probably the decrease of carbon dioxide during cooling. Thus, enflurane tends to induce an increase of EEG activity, which is closely correlated with an increase in cerebral oxygen uptake as demonstrated elsewhere [40]. The addition of CO_2 to the anesthetic mixture is often advocated to prevent such episodes of spike-like patterns [41]. These changes brought about in the EEG by enflurane differ from other agents. *Hypercarbia*, as it deepens anesthesia, may also function as a form to prevent a spike-like pattern in the EEG. If burst suppression is present, the suppression pattern is lengthened and the burst shortened as CO_2 is introduced. But during enflurane anesthesia, *hypocarbia* lengthens bursts and shortens the suppressions. Thus hypocarbia can be considered as an adjunct to enflurane which potentiates the initial central effects of the inhalational agent while the epileptiform components are enhanced [42, 43].

As with the other volatile agents, addition of N_2O results in an increase of the anesthetic effect.

4.3 Intravenous agents and their effect on the EEG

Barbiturates. The injection of a short acting barbiturate initially produces fast 15–30 Hz waves. Thereafter 5–12 Hz waves superimpose themselves on the fast waves often in spindle-shaped bursts. Next, high-amplitude 1–3Hz waves develop indicating a depth of anesthesia at which skin incision is tolerated (Fig. 23). If the barbiturate is given rapidly in a bolus, the earlier stages may be by-passed and delta waves will first appear. Increasing the dose of the barbiturate results in burst suppression in which the suppression period progressively is lengthened as the dosage is increased until electrical silence occurs.

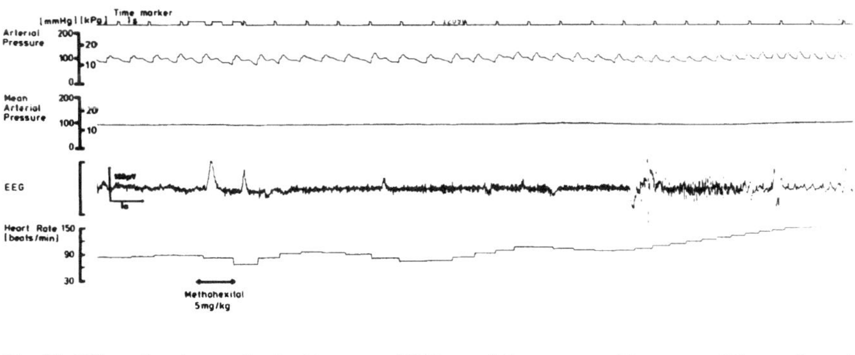

Fig. 23. Effect of a short acting barbiturate on EEG, arterial pressure and heart rate. 15 seconds post injection slow delta to theta waves develop while, simultaneously, heart rate increases. The latter is considered to be related to a desinhibition of brain stem structures involved in cardiovascular control.

It is suggested that barbiturates reversibly block the synchronizing influence of lower brain stem structures, which results in a preliminary EEG arousal with fast activity and an increase in heart rate. As the barbiturate reaches midbrain structures and the rostral parts of the encephalon, a synchronized pattern results dominated by spindles [44].

Etomidate. The induction dose of 0.3 mg/kg results in overlapping four stages. First there is an increase in alpha amplitude followed by a mixed theta activity. Second there is mixed theta to delta activity while in the third stage consistent delta waves are apparent. In the fourth stage burst suppression appears [45]. Myoclonic and tonic movements, resembling those of seizures in some patients, are *not* associated with seizure activity in the EEG [45].

Ketamine. After a normal dose of 1–2 mg/kg of ketamine the EEG consists of a mixture of 5–6 Hz, moderate high amplitude activity with very fast 30–40 Hz short amplitude waves superimposed. With high dosages such as 2–4 mg/kg, spike wave complexes can occur which, with further increase, result in generalized seizure activity [46, 47].

Opioids. While opioids, as they are used for premedication, have little effect on the EEG, high doses as they are used in techniques for cardiovascular surgery, result in a fall in frequency and in an increase in amplitude. The resultant dominant delta activity is demonstrated in a patient scheduled for elective cardiovascular surgery, who received 0.3 mg of the potent opioid fentanyl for the induction of anesthesia (Fig. 24).

Such effects are not only associated with fentanyl, they are also seen after the short-acting narcotic alfentanil, high doses of morphine (1–10 mg/kg), and after moderate doses of sufentanil (15–30 μg/kg) [48–51].

With increasing concentrations of the opioid at the receptor site in the CNS, delta waves become dominant. This effect of CNS-changes is also reflected in a drop of heart rate signalling a close correlation between EEG changes and the centrally mediated drive of the heart rate which results in bradycardia generally seen after opioid administration (Fig. 25).

Such central effects of opioids on EEG activity can be reversed by the specific antagonist naloxone resulting in a dominance of beta activity and an arousal reaction (Fig. 26).

Pethidine, morphine, alfentanil, fentanyl, and sufentanil in doses higher than 20, 180, 5, 4, and 4 mg/kg respectively have been shown to produce convulsions in the canine [53]. As such massive doses are never used in man, no such effects will be observed.

The clinical effect of the short acting opioid alfentanil on the EEG, determined to evaluate the onset of opioid activity when compared with fentanyl, can be represented in the spectral edge frequency of EEG power spectra. Fig. 27 shows the time course of the spectral edge (=degree of slowing of the EEG) and the corresponding concentrations of alfentanil (left) and those of fentanyl (right) in

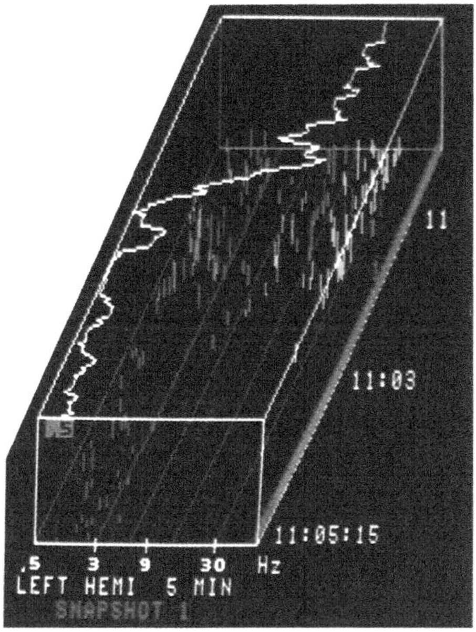

Fig. 24. Effect of 0.3 mg of fentanyl on EEG power spectra in a premedicated patient for the induction of anesthesia. Shortly after the injection there is a marked decline of fast beta- and alpha- and a dominance of delta-activity followed by a reduction in the activity edge as low as 0.5 Hz. The latter indicates the point at which 95% of all the electrical activity resides. Activity of the brain as it is analyzed and processed by the Lifescan^{TM} computer and then displayed in one grid. Amplitude gain 200 μV.

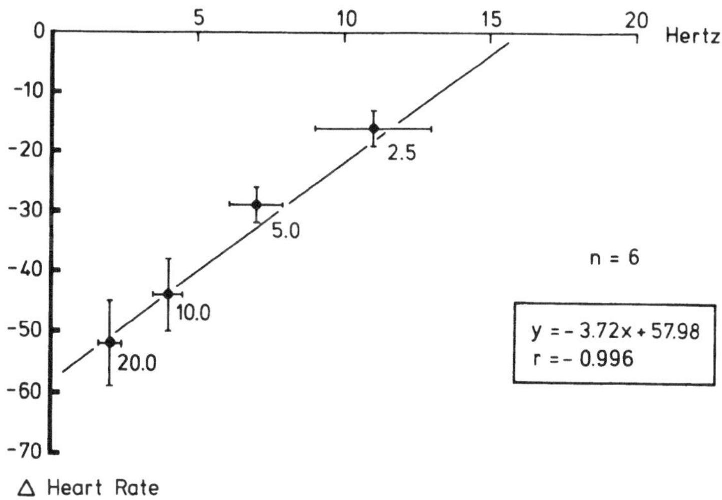

Fig. 25. Continuous perfusion of increasing concentrations of fentanyl 2.5–5.0–10–20 μg/ml through the IVth cerebral ventricle in the awake canine results in a dose-related decline of EEG frequency (adapted from Freye & Gupta [52]).

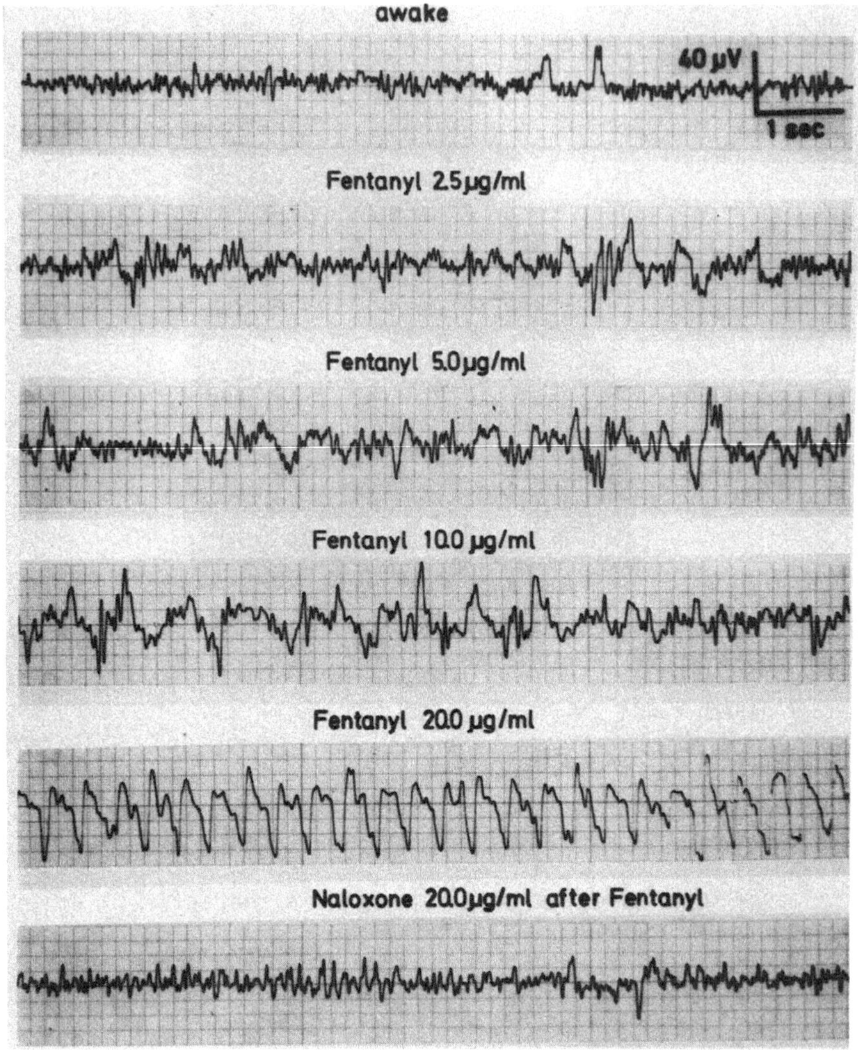

Fig. 26. The dose-related increase in slow wave activity of the EEG during fentanyl perfusion of the IVth ventricle is reversed by the perfusion of naloxone (adapted from Freye & Gupta [52]).

patients receiving either alfentanil 1500 µg/min or fentanyl 150 µg/min. With alfentanil the onset of effects was 1.1 min, while spectral edge closely correlated with serum concentration. With fentanyl onset of effects was 6.45 min with spectral edge changes lagging behind serum concentrations [54].

Both, alfentanil and fentanyl cause similar EEG changes, characterized by progressive slowing in frequency and increase in amplitude. Maximum effect is characterized by delta waves of slow frequency and large amplitude.

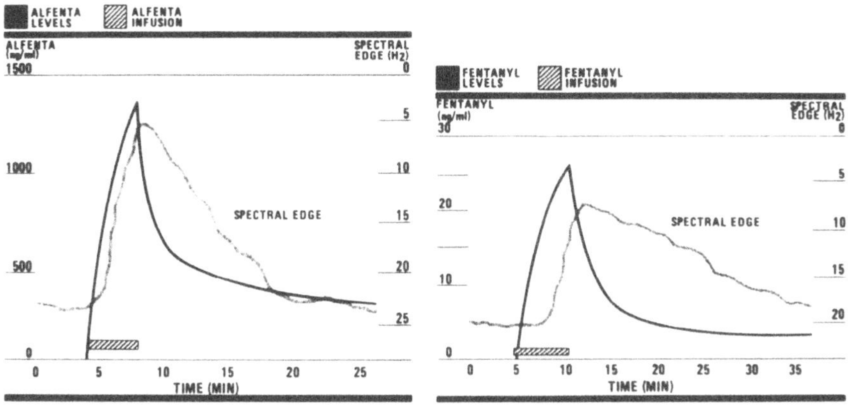

Fig. 27. Change of spectral edge frquency and of serum concentration after fentanyl and alfentanil-infusion respectively. The faster decline of spectral edge frequency after alfentanil reflects a faster elimination of the compound from the CNS and a faster recovery from the anesthetic. Note, the close correlation of spectral edge and plasma alfentanil-levels (adapted from Scott et al. [54]).

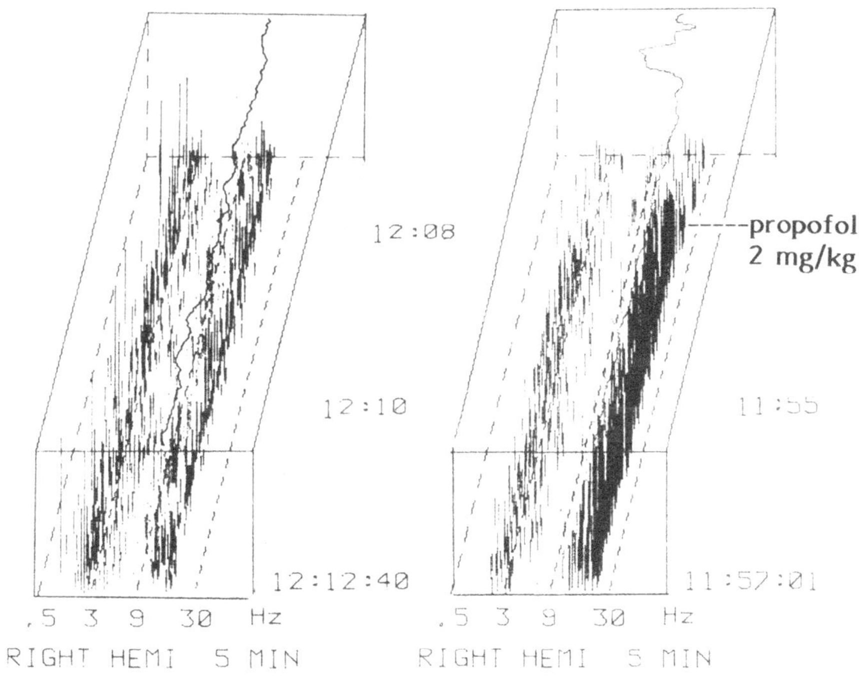

Fig. 28. Characteristic power spectra and the corresponding spectral edge in the fast alpha to beta range as a result of the injection of midazolam 15 mg (a, left) and propofol 2 mg/kg respectively (b, right) used for the induction of anesthesia. When N_2O is added the latter results in a shift of power in the lower frequency range.

Although the reliability of the EEG as a measurement for correlation of dose with clinical effect has not yet been established, its potential use has thus been demonstrated, showing a relationship between plasma levels and drug effect [55, 56].

Benzodiazepines. Benzodiazepines in a dose as they are used for preoperative anxiolysis have some effect on the EEG which is characterized by an increase in alpha activity (=synchronization) with resultant sedation. However, after the use of high doses as for induction (midazolam or diazepam) the EEG, *contrary* to all the other induction agents, shows an increase in fast EEG activity of the beta-band which reaches its maximum when the patient is asleep (Fig. 28a). Such an increase of beta in the EEG is a characteristic effect of all benzodiazepines, those with a more hypnotic component also will accentuate delta-activity [57].

Propofol. Similar effects to those of benzodiazepines can be observed when the hypnotic propofol is given for the induction of anesthesia (2 mg/kg). The increase of activity in the beta range of the EEG which coincides with unresponsiveness of the patient suggests a similar effect for benzodiazepines and propofol (Fig. 28b). Due to its high lipophilicity, the compound rapidly passes the blood brain barrier and induces its central effect. Thus, when benzodiazepines and/or propofol are used, the depth of anesthesia *does not* correlate at all with the shift of power into the higher frequency range. However, once N_2O is added, the maximum power is shifted to the alpha band [58].

CO_2. The effects of CO_2 on the EEG vary according to the $PaCO_2$ level. While 30% inhaled CO_2 result in an intermittent epileptiform discharge, higher concentrations of 50% produce electroencephalographic depression in the animal. Hypercarbia at a steady halothane-level and contrary during enflurane anesthesia results in slowing of the EEG [59].

5. The EEG and cerebral ischemia

At this time the most significant contribution EEG monitoring can make to patient care, is that of the early detection of cerebral dysfunction, especially that due to ischemia. Several possible causes of reduced cerebral oxygen delivery are shown in Table 2.

The functioning of the cerebral cortex is extremely sensitive to any PaO_2 changes. Insufficient cerebral blood flow or inadequate partial pressure of oxygen is reflected within seconds in the EEG (Fig. 29) [60–62].

Table 2.

Factors reducing oxygen delivery to the CNS

1. Decreased oxygen-carrying capacity:
   - Anemia
   - Hypovolemia
   - Carbon monoxide poisoning
2. Decreased arterial oxygen content:
   - decreased inspired oxygen (hypoxic mixture)
   - inadequate ventilation (disconnection, malfunction)

Factors decreasing cerebral perfusion pressure

1. Decreased systemic arterial pressure:
   - low blood volume
   - low cardiac output (pump failure, aortic clamping, arrhythmias)
   - blood pooling (high dose opioids, nitroprusside, nitroglycerin)
2. Increased intracranial pressure:
   - hypercapnia
   - acute systemic hypertension
   - mass lesions (hematoma, tumor, ruptured aneurysm)
   - post head injury with edema
3. Mechanical obstruction of cerebral vessels
   - thrombus
   - embolus
   - air (on occasion)
   - surgical clamping of carotid artery
   - vasospasm

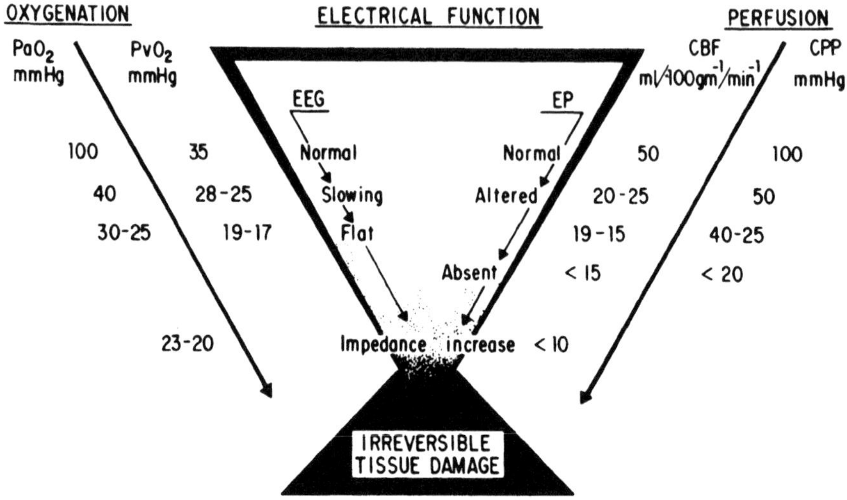

Fig. 29. The changes in brain electrical activity after the decrease in oxygenation of blood, the decrease in cerebral blood flow (CBF), and cerebral perfusion pressure (CPP).
PaO_2 = arterial oxygen tension; PvO_2 = mixed venous oxygen tension.

The typical pattern in the EEG seen during cerebral ischemia is:

1. reduction or loss of high frequency (Fig. 30),
2. appearance of large amplitude slow waves in the delta range, and
3. prolonged ischemia eventually results in a decrease of amplitude and frequency leading into isoelectricity.

Even a local decrease in cerebral blood flow, as it occurs in cerebral embolism, results in an EEG change which becomes visible in spectral analysis of the EEG recorded some distance away from the anatomical site of the ischemia [60]. This is mainly due to the fact that computer analysis is *more sensitive* to changes in high frequency pattern than normal interpretation.

The EEG appearances of any ischemic episode are similar. Differentiation may be best made by closely observing the clinical situation focussing on:

a) blood pressure,
b) ECG,
c) oxygen saturation,
d) surgical manoeuvers, and
e) drugs being administered.

Other concurrent factors which alter the intraoperative EEG are:

f) change in the depth of anesthesia,
g) temperature changes (controlled hypothermia), and
h) changes in CO_2 content.

The latter factors, however, may be recognized by their *relatively slow onset* lasting for several minutes as opposed to the changes of ischemia which generally occur within seconds. One should always keep in mind that there are situations where the

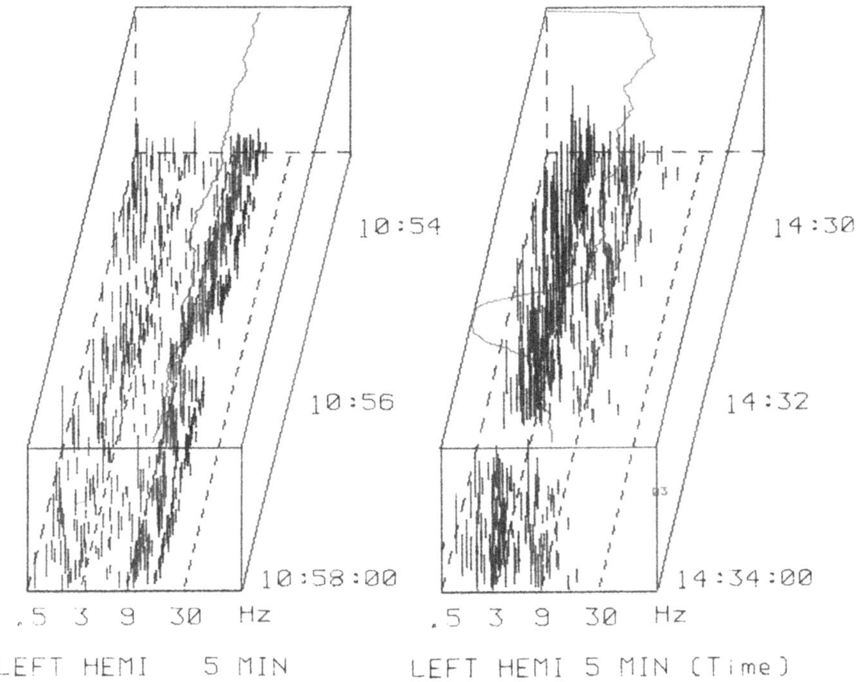

Fig. 30. Two representative examples of patients on ECC (extracorporeal circulation) where temporary stop of the pump for only three seconds results in a temporary loss of fast EEG activity.

EEG may occur acutely depressed or ischemic upon injection of an anesthetic, which rapidly can pass the blood brain barrier. Such situations may be found in high-dose opioid anesthesia, where fentanyl induces an immediate and marked reduction of fast frequency of the EEG with resultant low frequency/high amplitudes in the delta range (Fig. 31).

Thus, attention should always be put to the clinical context during which the EEG is observed. Also, it should be noted that a decreased activity with increased amplitudes pertains particuarly ischemic events of the cortex, whereas ischemia of the capsular interna or the thalamus may result in unremarkable EEG changes or alterations of the alpha-rhythm in awake patients respectively. Brainstem ischemia however, appears in the EEG as a widespread slowing and a decrease in amplitude [63].

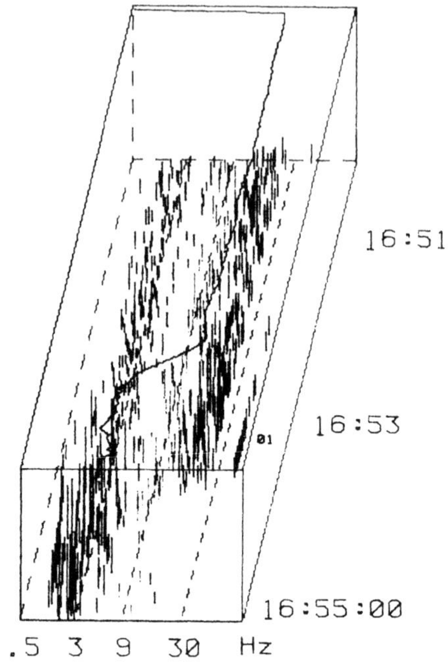

Fig. 31. Example of the sudden drop of EEG activity in a patient who had just received 0.35 mg of fentanyl for the induction of anesthesia for CABG (event No 01). If no anesthetic would have been given such a drop is characteristic for malperfusion.

5.1 Differentiation between ischemia and anesthetic-induced EEG changes

As such a differentiation is of clinical importance, some of the major points outlining the induced changes in the EEG will be summarized as follows:

1. Changes are of *focal nature*. As such changes are seldom anesthetic-induced, they rather relate to intracerebral or cerebrovascular pathology. The situation may, however, become more complex when global hypoxia has to be differentiated from a deepening of anesthesia or hypothermia. Yet, the clinical situation and the pattern of the original EEG wave provide important information allowing the determination of the etiology of the EEG changes.

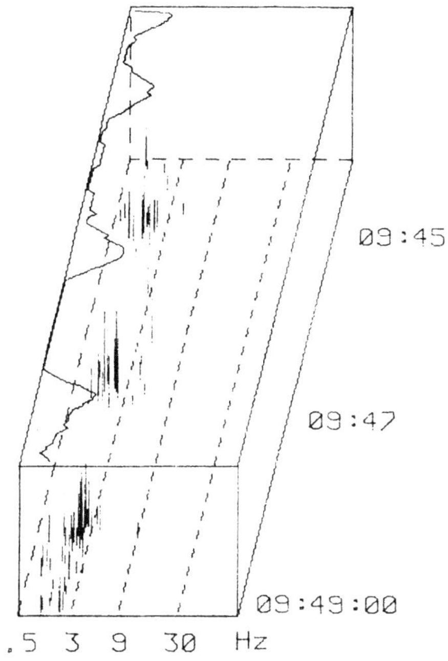

Fig. 32. Representative example of a patient receiving isoflurane-oxygen anesthesia while being on ECC. Mean perfusion pressure 50 torr. With a core temperature of 25 °C, an isoelectric pattern with intermittent slow delta waves develops. The latter is clearly seen in the fluctuating activity edge (EEG power spectra derived with the Lifescan™ cerebral monitor).

2. *Severe hypoxemia* produces an *isolectric* (flat-line) EEG. A similar pattern may also be produced by deep hypothermia (Fig. 32), large doses of barbiturates greater than those used to induce anesthesia, and high isoflurane concentrations. It is, however, conceivable that neither hypothermia nor barbiturate injection of sufficient magnitude could occur without the awareness of the anesthesiologist.

3. *Onset of EEG slowing.* The pattern of development of EEG slowing is important as inadequate cerebral oxygen supply is abrupt, thus slowing the EEG, while deep anesthesia requires time to develop. Thus, the onset of EEG slowing is of a more gradual nature. Hyperventilation may also produce EEG slowing. This, however, is in response to cerebral vasoconstriction as induced by hypocarbia which in return produces hypoxia of a mild degree. These changes are also abrupt in onset and the coexistance of hypocarbia as an etiology can be determined by blood gas analysis or end-tidal carbon dioxide determination.

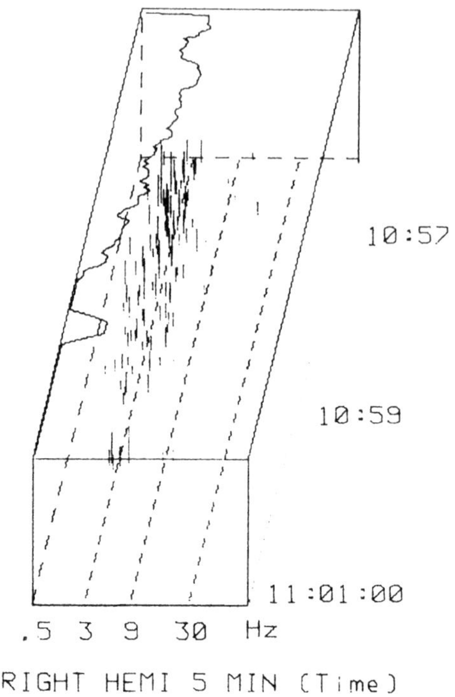

Fig. 33. Representative example of a patient undergoing AC by-pass operation under isoflurane anesthesia. Body core temperature 31 °C, and mean perfusion pressure 50 torr.
In the EEG there is a gradual loss of activity lasting over 2 minutes. Only when perfusion pressure was raised by the pump technician EEG activity in the delta band reappeared. The patient uneventfully recovered from the operation. Time scale on the right hand side.

4. *Addition of cardiovascular changes* (usually hypotension). In the presence of excessive inhalational anesthesia this may occur before, or together with, the EEG changes (Fig. 33). The change in cardiovascular parameters then provide information of the etiology of EEG slowing. Additionally, enflurane and isoflurane produce burst suppression prior to or with EEG slowing. Ischemia *normally* does not produce burst suppression.

5. *High dose opioids* may also produce an *abrupt* onset of EEG *slowing*. This would be likely to occur at the start of the case and under a situation in which question about the etiology of the EEG change is unlikely. Severe hypoxia during opioid anesthesia still produces an isoelectric EEG.

In summary, the EEG signal by itself *does not* allow the determination of the etiology. It is the same as measuring a blood pressure of 80/50 which does not tell whether this drop is produced by volume loss or vasodilatation. Thus, it is important to *simultaneously* consider the clinical setting, as in blood pressure recording, when the EEG is recorded. Only then is differentiation of deep anesthesia from hypoxia in EEG monitoring possible.

5.2 Cerebral monitoring during hypothermia and extracorporeal circulation (ECC)

The greatest potential for ensuing CNS damage is that during ECC. This is due to a number of causes:

a) sudden hemodilution with the onset of ECC,
b) change from pulsatile flow to non-pulsatile flow,
c) reduction in mean arterial pressure (see Fig. 32),
d) exposure of blood to foreign material,
e) an embolic event by loosening of arteriosclerotic material during manipulation at the great arteries, and
f) air embolism. [64]

The major reduction of body temperature by itself, although planned, has a dramatic effect on the EEG. Even at stable concentrations of the anesthetic, hypothermia produces progressive slowing in the EEG frequencies in a step-related fashion. Ultimately, there is a reduction in amplitude followed by intermittent suppression, and finally total suppression of all EEG activity which occurs with rapidly developing hypothermia around 25 °C [65].

A representative example depicts the various changes the EEG power spectrum will undergo during hypothermia. In spite of the continuous steady level of isoflurane via by-pass into the oxygenator, there is a gradual decrease of frequency and of spectral edge activity (solid line), as viewed in a typical Lifescan™ display on screen. In the 'glass box', results are displayed in a three-dimensional moving picture with the various EEG frequency bands at the X-axis, their corresponding amplitudes at the Y-axis while the Z-axis into the 'depth' of the screen reflects time (Fig. 34).

Note that with deeper body core temperature burst suppression will occur which tends to last longer until finally there is a flat processed EEG (Fig. 34a). These changes are reversed as body core temperature goes up again, resulting in an increase in amplitude and frequency of the EEG. The latter is signified by a shift of the 'spectral edge activity' from the left (delta band) to the right (alpha band) (Fig. 34b).

5.3 EEG monitoring during carotid endarterectomy

The relevance of changes in the EEG to neurological outcome after carotid endarterectomy has been challenged [66]. Reports have outlined the fact that the appearance of transient EEG abnormalities during operation does not necessarily imply a postoperative neurologic dysfunction. If intraoperative changes lasted longer than *20 minutes*, inevitable consequences would result [60]. The same may hold true in the use of EEG monitoring to determine adequacy of cerebral perfusion during cardiopulmonary by-pass. Even though prolonged isoelectric EEG activity secondary to ischemia does not imply irreversibility, it does, however imply evidence of ischemia and an immediate danger of tissue damage, one which should

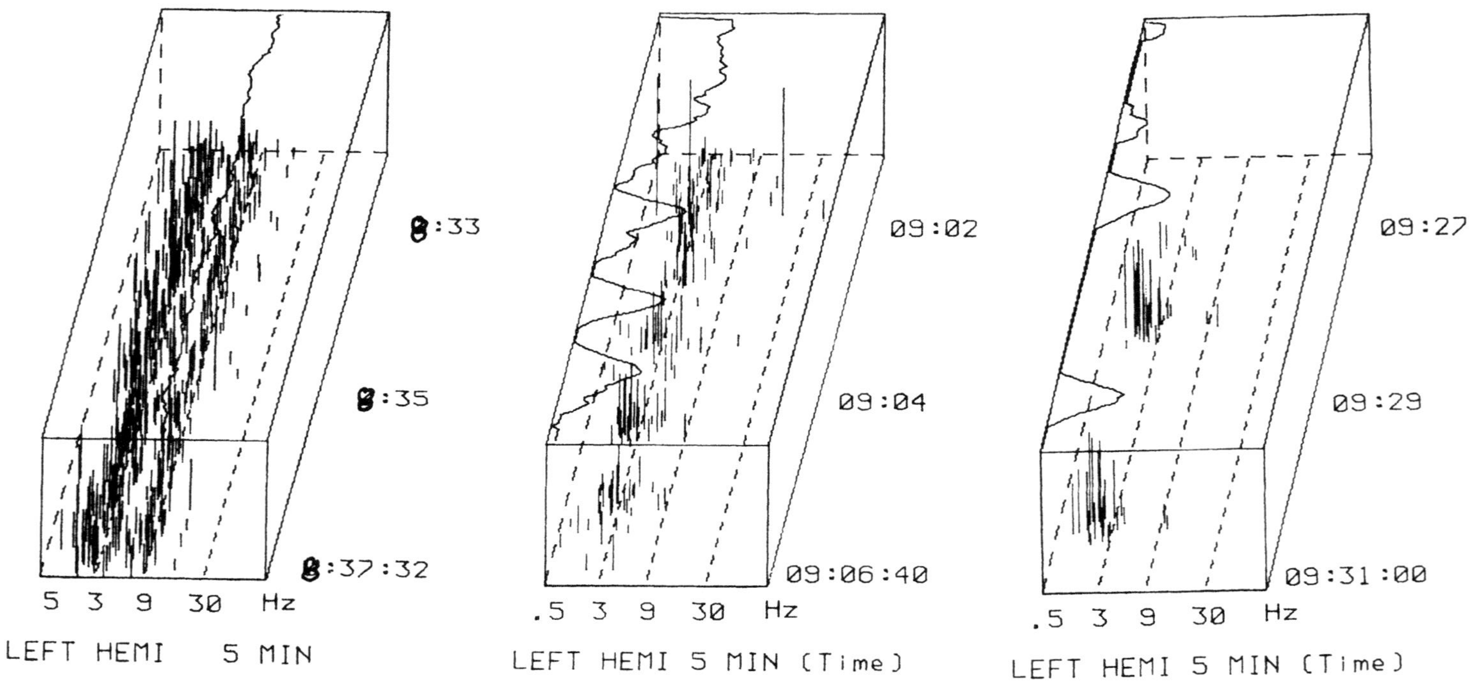

Fig. 34. Change of EEG power spectra and activity edge followed by a drop of core temperature from 36 °C to 22 °C inducing periods of electrical silence in a patient on ECC during isoflurane anesthesia (1 Vol%).

a) Hypothermia: *left* 35 °C, *middle* 27 °C, *right* 22 °C.

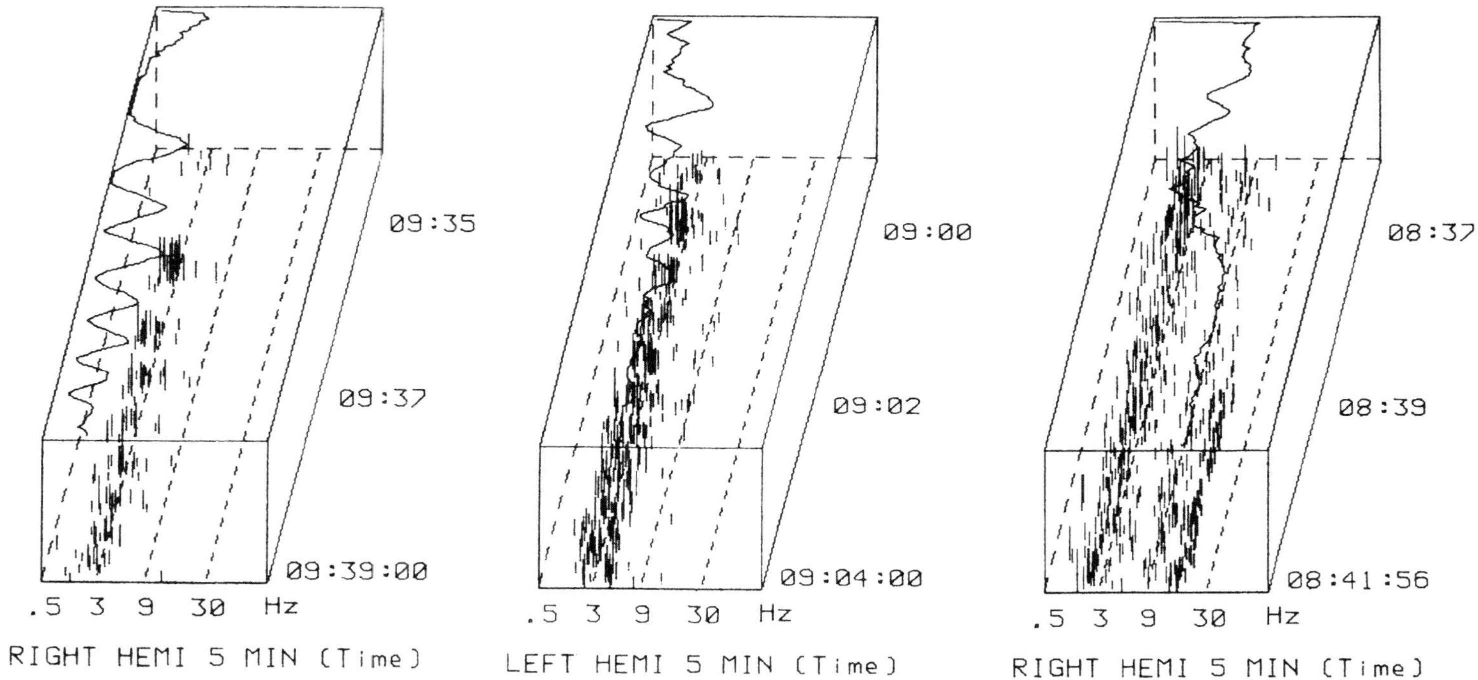

Fig. 34. Change of EEG power spectra and activity edge followed by an increase of core temperature from 25 °C to 36 °C inducing an increase of fast-activities in a patient on ECC during isoflurane anesthesia (1 Vol%).

b) Rewarming: *left* 25 °C, *middle* 30 °C, *right* 36 °C.

be addressed immediately. Subtle complications such as *impairment of memory* may occur after ischemic episodes of less than 10 minutes. On the other hand, total cessation of cerebral perfusion at hypothermia of 18 °C and an isolectric EEG will result in *no damage* unless this isoelectric episode does not exceed 30 minutes [67, 68]. In carotid endarterectomy the following EEG changes suggest insufficient cerebral blood flow (Fig. 34c):

1. With the onset of clamping there is a sudden slowing of activity.
2. The drop of EEG activity is related to the side of clamping (side-difference).
3. Once the artery is declamped fast activity of the affected hemisphere will reappear (if there was no prolonged period of clamping).

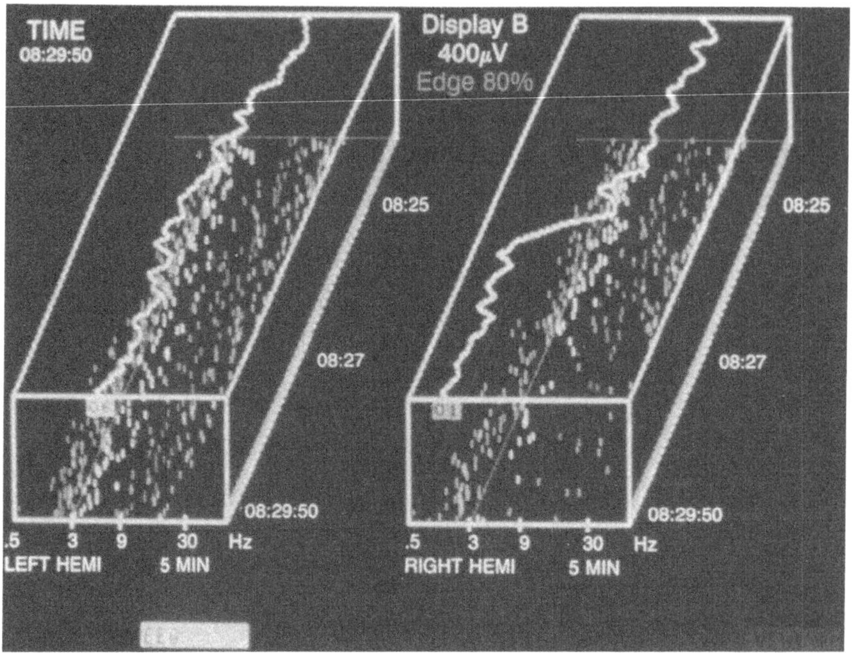

Fig. 34c. A typical screen from the Lifescan™ brain activity monitor. On the right hand side there is a sudden shift of the activity edge to the left indicating a decrease on energy while the edge on the left remains constant. A clamp has been placed on the right carotid artery, thus decreasing the blood flow to and the activity of the right side of the brain. If the situation is left unattended, permanent brain damage will result.

6. Cerebral monitoring in the intensive care unit

6.1 Introduction: Representative case reports

Cerebral monitoring may also be used in the intensive care unit for the same reason as in the OR to which the following situations are added:

a) as a monitor for epileptic activity in the unconscious patient,
b) as a dosage guide in the management of barbiturate coma,
c) as a dosage guide in the management of analgosedation,
d) as an objective measure in the management of drug overdose,
e) as a screening monitor for situations in which 'brain death' is suspected, and
f) as a continuous display of cerebral function similar to that of the continuous monitoring of cardiovascular parameters (e.g. arterial blood pressure, central venous pressure, pulmonary artery pressure, pulmonary wedge pressure, etc.).

The following section presents a collection of computerized EEG power traces illustrating some of the applications of EEG monitoring in the ICU. All cases were monitored with the Lifescan™ Monitoring System (Neurometrics, San Diego, USA). EEG power spectra were usually derived from position F_3 and F_4 with the corresponding negative electrode placed on the mastoid. For grounding the electrode was positioned on the forehead (FpZ).

6.1.1 Epileptogenic activity

Patient with obliteration of the left, and narrowing of the right carotid artery after resuscitation (Fig. 35).

In the EEG power spectrum repetitive seizure activity over the right hemisphere with paroxysmal discharge is reflected in an increase of power in the beta band. As no tonic-clonic muscle activity was visible, epileptogenic discharges could only be made obvious with the aid of the cerebral monitor. The EEG power spectra were useful in several respects:

a) to localize the site of origin of epileptic foci (asymmetry between both hemispheres),

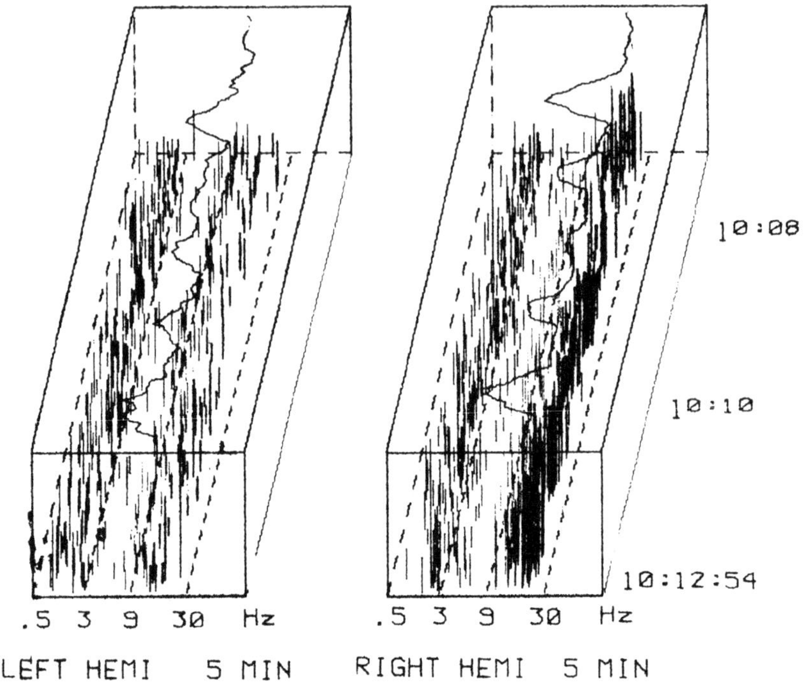

Fig. 35. A 68 year old male after resuscitation with right sided epileptogenic activity. In the EEG power spectra fluctuating power peaks in the beta domain are evident. There is a spread of epileptogenic activity to the left hemisphere. The effectiveness of antiepileptogenic agents can be monitored.

b) to evaluate the effectivity of the therapeutic regime,
c) to select the individual dose, necessary to reduce seizure activity, and
d) as a follow-up over the next days.

Over the left hemisphere the spread of seizure activity (i.e. increase of power in the beta band) is reduced which suggests that the source of seizures is on the right hemisphere.

6.1.2 Monitoring in head trauma

Another case shows the EEG power spectra of a 42 year old patient involved in a car accident suffering from multiple fractures of arms and legs, with a third degree head trauma. Partial tetraparalysis with flaccid paralysis of the left arm was evident. Computer tomography (CT) revealed no signs of intracerebral or subdural mass bleeding. The power spectra showed a pathological slowing of EEG waves over the right hemisphere. Over the left hemisphere fast activity in the beta domain was seen, reflecting a fluctuation of vigilance. The latter is visualized by changes in the spectral edge. The advantage of computerized cerebral monitoring for such for such a case is obvious when answers to the following questions are contemplated:

a) Is there an interhemispheric difference of activity?
b) Is there any deterioration of cerebral activity over the past 24 hours?
c) What is the effectiveness of the therapeutic regime?
d) Has there been any sign of epileptogenic activity which demand an early intervention with antiepileptogenic compounds?

6.1.3 The EEG post cerebral malperfusion

A 75 year old male developed hypovolemia postoperatively after desobliteration of the femoral artery. His medical history reveals a multiple vessel sclerosis with transient ischemic attacks (TIA). He is in a comatose state. On nociceptive stimulation synergism of the left arm is more pronounced than on the right side.

In the EEG power spectra a marked reduction of alpha to beta activity over the right fronto-temporal area is obvious. Pathological slow waves in the delta-range are dominant (Fig. 36). Due to the past history, the incidence of acute hypovolemia and the present slow waves, encephalopathy is diagnosed with signs of a right sided decerebration.

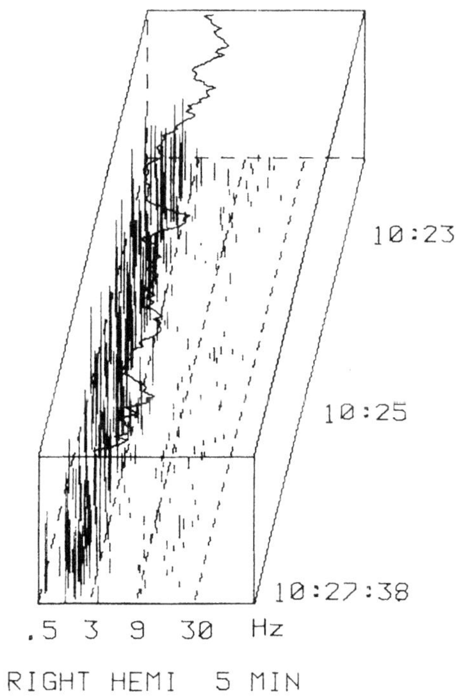

Fig. 36. Sequelae of acute malperfusion in a patient with previous borderline cerebral perfusion. Marked right-sided pathological slow waves suggest damage of cortical structures.

The latter case demonstrates that, in addition to neurological exploration, the EEG power spectra are helpful in many respects:

a) they provide insight information of pathological EEG activity of the whole cortex,
b) give some idea in regard to the prognostic outcome, and
c) will show up any sudden deterioration in cortical activity.

6.1.4 Sedation in an ICU setting

One of the major concerns in the intensive care situation is the question of how to sufficiently sedate the patient who is on the ventilator, avoiding any overhang of sedatives when the patient is ready to be weaned off. In such a situation the EEG power spectra may become useful as they give an idea about the degree to which sedatives depress the cortex. Deep sedation usually is accompanied by a loss in fast activity in the beta and alpha domain. Doses can thus be tailored individually in order to guarantee a sufficient sedation and at the same time avoid a potential overdose. As most regimes, which are used for analgosedation in the ICU, cannot be judged by cardiovascular depression, it seems only logical to look into that organ which is considered the target of any sedative, the CNS.

Fig. 37 shows the mean effect of analgosedation using the short acting compounds alfentanil and midazolam administered via a perfusor to patients who are in need of artificial ventilation. It can be seen that high perfusion rates induce a

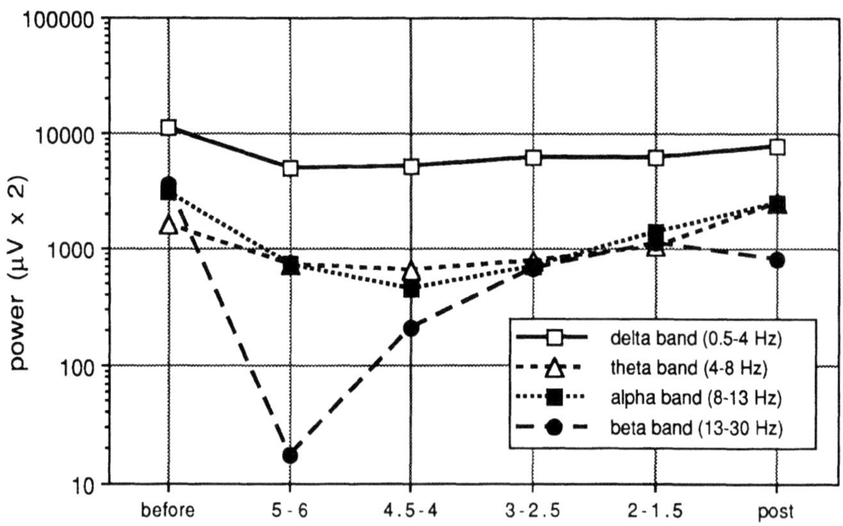

Fig. 37. Change of power in the various frequency bands during the administration of different amounts of an alfentanil/midazolam mixture given to patients (n=10) for analgosedation in an ICU setting. Note that the fast frequency beta-band reflects the dynamics of both drugs on CNS activity.

depressive effect on fast power spectra. The close correlation of power in the fast beta and alpha domains is evident with a reduction in dose. Thus, EEG monitoring during any type of analgosedation:

a) is a useful tool to guide the clinician in adjusting to the individual dose necessary,
b) is a way to avoid the incidence of drug overhang,
c) immediately notifies the clinician on the efficacy of the drugs used for sedation, and
d) gives early warning on the possible development of tachyphylaxis during long-term administration.

6.1.5 EEG power spectra post mitral valve replacement

Use of cerebral monitoring in the ICU for the preservation of cerebral function is outlined in the following case: A patient in the ICU who had undergone mitral valve replacement, in spite of anticonvulsant therapy still experienced seizure activity with clonic movements of the left side. Due to clonus of the left leg, artifacts are marked by the computer analyzing the EEG waves (Fig. 38). The seizure activity was of short duration spreading over to the left hemisphere. At 11.12 a.m. (event No. 03) clonus is evident; on the print-out the development of fast activity with high power is first seen on the right hemisphere and later spreads to the left side. Additional therapy with a barbiturate sufficiently depressed muscular jerks. However, in the EEG, high power in the beta domain still reflects abortive seizure activity. Thus, monitoring for the evaluation of effectivenss of drug therapy is very useful. If the patient had not been monitored on a continuous base, abortive seizure activity would not have been noted. As continuous seizure activity results in an increase in cerebral metabolic rate and an increase in cerebral oxygen consumption, damage of border-line nerve cells might ensue.

6.1.6 EEG power spectra and focal seizures

Focal activity of seizures may look different when a number of anticonvulsants have already been administered, which however do not completely abolish hyperactivity of a circumscribed cortical area. This is demonstrated in another patient who after mitral valve replacement developed seizures. With appropriate treatment seizures completely disappeared. In the EEG power spectrum, however, seizure activity is still apparent in the right hemisphere (Fig. 39). The hyperactivity is visualized in the 'glass box' derived by the Lifescan^{TM} cerebral monitor where a fluctuation of power is reflected in the spectral edge. No such increase in spectral edge is seen in the power spectra of the left hemisphere. Insufficient anticonvulsant therapy was diagnosed and the dose increased as indicated by EEG monitoring. Thus, EEG power spectra have helped in the evaluation of sufficient dosing in that particular patient. In addition, the EEG is:

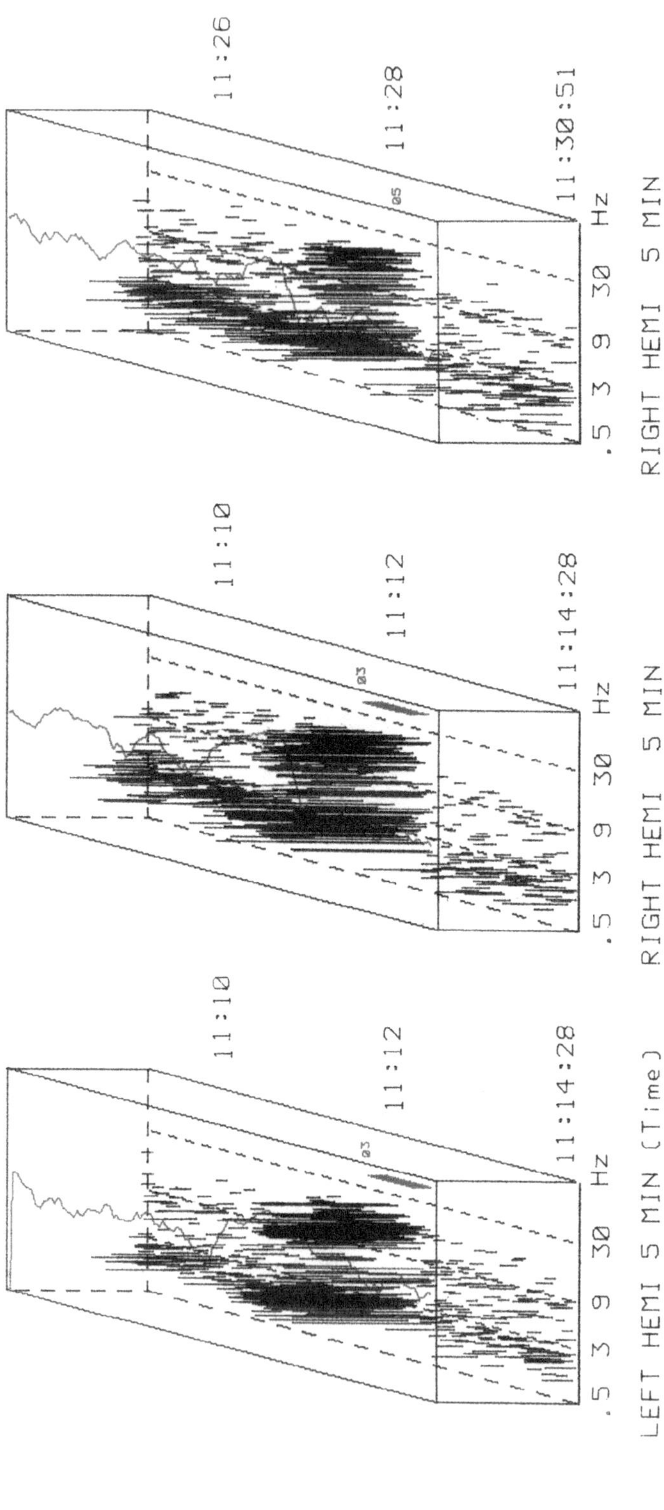

Fig. 38. EEG power spectra of a patient developing seizures after mitral valve replacement. Although the additional injection of a barbiturate resulted in cessation of tonic-clonic movements, the EEG power spectra still show seizure activity, suggesting insufficient dosing.

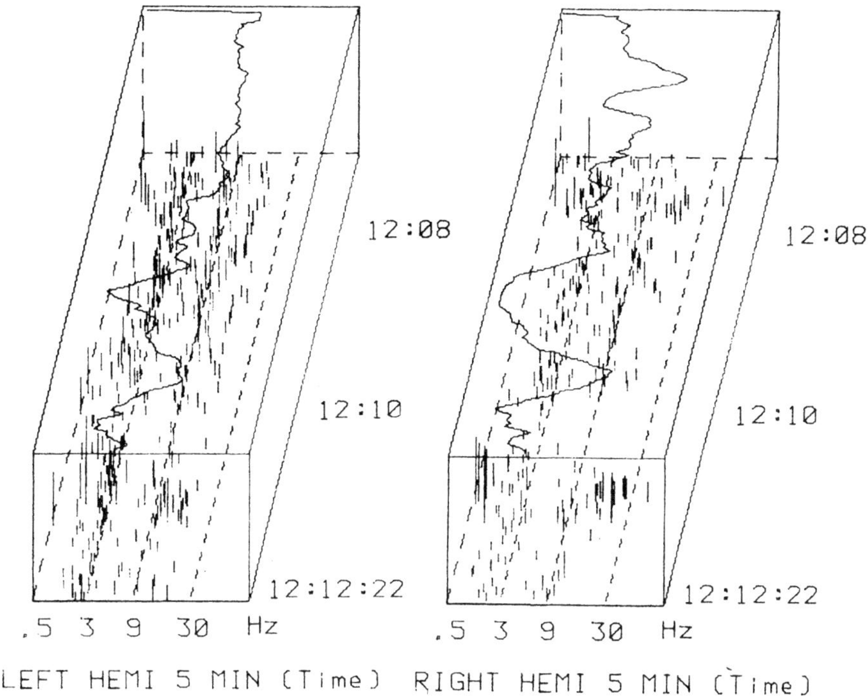

Fig. 39. Another patient after mitral valve replacement with general seizure activity where anticonvulsant therapy induces a suppression of the hyperexcitatory state of the cortex. An active focuse however, can be derived from the marked fluctuation of the activity edge in the chrono-power spectrum.

a) an indicator when to reduce the dose,
b) an indicator when to increase a dose,
c) able to indicate the ineffectivity of a specific drug,
d) able to indicate the efficiency of a combination of drugs, and
e) an aid in evaluating the severity of seizure activity in the long run.

6.1.7 EEG power spectra and benzodiazepine intoxication

Another case reflects the advantage of cerebral monitoring in the emergency and/or detoxification unit where comatose and unconscious patients, due to drug overdose, are treated. The brain as the prime target of centrally active drugs should be monitored similar to any other organ which is at risk in such a critical situation.

A 28 year old female was found in a comatose state in the downtown area. Due to several bruises and lacerations on her head, which raised suspicions of rape, she was taken to the emergency unit. Benzodiazepine intoxication was also suspected as an empty package of flunitrazepam (40 mg) was found in one of her pockets. Computerized tomography and careful neurological exploration revealed no signs

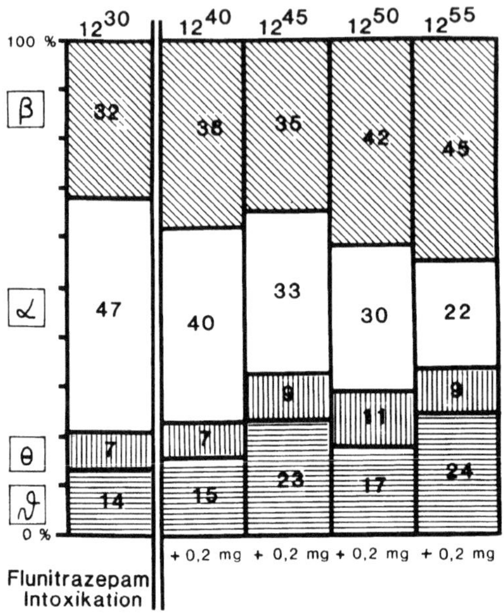

Fig. 40. Relative % changes in the various EEG power spectra in a comatose patient with benzodiazepine intoxication receiving incremental doses of the specific antagonist flumazenil (Anexate™). The increase in vigilance is reflected in an increase in beta-activity.

of an intracranial lesion. Blood gas analysis at that time was PaO_2 44 with a $PaCO_2$ of 46 mmHg. Due to deterioration of respiration she was intubated and ventilated with oxygen. Blood gas analysis now was PaO_2 356 and $PaCO_2$ 42 mmHg. As intracarnial lesions still could not be ruled out and in order to evaluate cerebral activity, she was monitored with the Lifescan™ computerized EEG analyzer. EEG power spectra revealed a dominance of activity in the alpha and theta bands with little pathological delta waves (Fig. 40). As benzodiazepines in the EEG usually increase fast activity and with a suspected flunitrazepam overdose in mind, the specific benzodiazepine antagonist flumazenil (Anexate™) was given in incremental doses. The relative power changes, as they were derived from the computed power spectra, showed a marked increase of activity in the beta band following the fourth and the fifth incremental dose of the antagonist (Fig. 39). Power in the alpha band in the comatose state dropped from a relative value 47% to 22% and beta increased from 32% to a relative value of 45%. These CNS changes reflected the patients behaviour as she slowly regained consciousness, had sufficient spontaneous respiration, did not tolerate the endotracheal tube any more, and after being extubated was orientated to space and time responding appropriately to verbal commands. Blood gas analysis at this time was PaO_2 67.1 and $PaCO_2$ 39.6 mmHg with no gross changes in blood pressure and heart rate. While still awake, the patient was sent to the detoxification unit for further surveillance. A rebound of benzodiazepine effect had to be anticipated (shorter half life of the antagonist) and although the patient became drowsy in the next two hours, she could always be awakened. Ventilatory assistance was not necessary.

The case demonstrates that even in the emergency unit cerebral monitoring in certain cases not only is mandatory but may also aid in evaluating the underlying cause of a comatose state. Thus, the following advantages of cerebral monitoring in the above situation can be derived:

a) an additional, simple and fast aid in evaluating the underlying cause of a coma,
b) aid in the treatment of a suspected drug overdose,
c) an additional instrument to differentiate between an intracranial lesion and central drug effects in an otherwise comatose patient,
d) a warning signal for resedation and a possible deterioration of CNS function,
e) a valuable tool for monitoring in the detoxification unit for long term follow up, and
f) eliminating the need for ventilatory support.

6.1.8 EEG measurements to differentiate between drug overdose and brain lesions

The detoxification unit is not only a place where drug overdose has to be treated. In the intensive care, especially in cases with long-term sedation, where evaluation of the cerebral status is practically impossible, a drug like the benzodiazepine antagonist in addition to cerebral monitoring can be very useful. Such a case is of interest when the patient has to be ventilated and is sedated with benzodiazepines, and the time is up to differentiate between drug overdose and a potential post-traumatic brain lesion.

A 60 year old male with polytrauma and thoracic contusion, had received a total of 3 mg of lorazepam to tolerate postoperative ventilatory support over the past 36 hours. Cortical activity, as it was derived from the EEG power spectra (Fig. 41), showed a reduction of power in the fast beta band. In order to differentiate between a possible cerebral lesion and drug overdose, the antagonist flumazenil was given in a single dose of 0.2 mg. The antagonist induced a marked increase of power in the beta domain, indicating arousal. This increase in the fast EEG waves could be stabilized by repetitive injections of the antagonist so that the patient could be weaned off much faster from the ventilator.

This case demonstrates why cerebral monitoring is useful in an ICU setting namely:

a) in differentiating between drug overhang and a possible intracerebral lesion,
b) in adjusting incremental doses of the antagonist, just enough to demonstrate arousal, but not too much to induce a sudden awakening with sympathetic overdrive,
c) in titrating the amount of the antagonist in regard to cortical effects, and
d) in notifying a possible rebound of sedation which goes hand in hand with a decay in cerebral activity and a decay of other vital functions such as respiration.

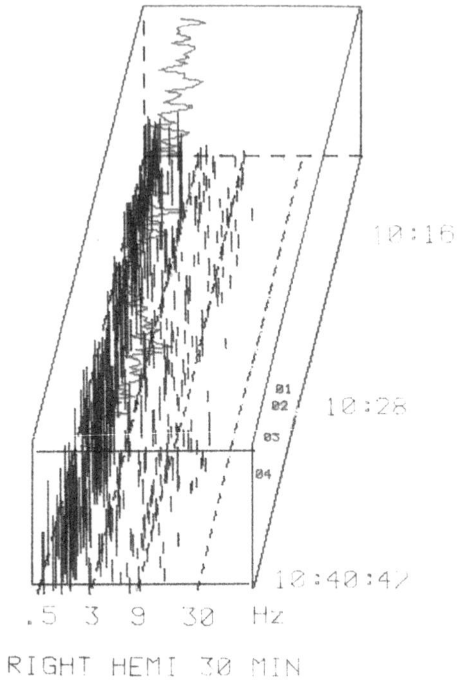

Fig. 41. Patient deeply sedated with a benzodiazepine on respirator therapy. In order to evaluate neuronal response, the benzodiazepine antagonist flumazenil is given. Shortly after the injection power in the beta domain increases suggesting viable cortical structures.

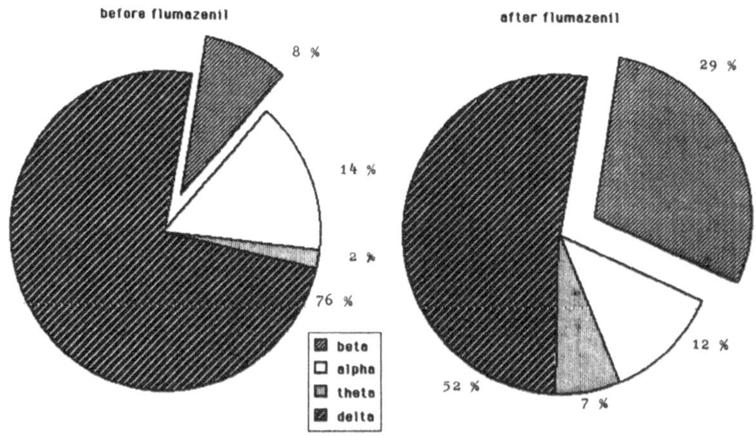

Fig. 42. Percent changes in EEG power spectra of patients in the ICU receiving benzodiazepines for sedation to tolerate ventilatory support (n = 10). In order to differentiate between primary brain lesion and drug overhang, the benzodiazepine antagonist flumazenil was given. The drug induced a marked increase in beta activity reflecting arousal. Additionally, the time necessary for ventilatory support could thus be reduced.

6.1.9 The 'diagnostic window' in long-term sedation

In certain cases during sedation of a ventilated patient, the use of a specific antagonists plus on-line EEG analysis is a valuable tool to evaluate cerebral function by 'creating a diagnostic window' into the sedative regime *without* total reversal of sedation. Such a 'diagnostic window' is characterized by an increase in fast activity in the beta and alpha domain in the sedated patient. A 'window' is present in Fig. 42 when deep sedation is partially reversed with the aid of the specific antagonist. As the half life of that compound is of short duration, sedation becomes dominant again after a few minutes, suggesting a reactive and not severely traumatized cortex. This reflects one of the few techniques to estimate cerebral activity, a technique which otherwise is impossible in a sedation regime.

6.2 EEG power spectra to differentiate between severe head trauma and drug overdose

Thus, the pharmacological tool in conjunction with automated EEG analysis is useful in all cases of inadvertent overdose of a sedative. Fig. 43 represents the case of a 22 year old male who had been involved in a motorcycle accident. During

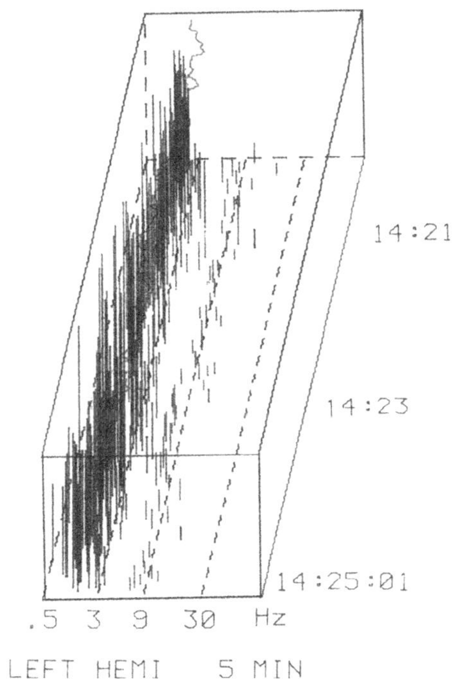

Fig. 43. Power spectra in a patient after severe head trauma. Although the patient had a suspected overhang of diazepam, pathological slow waves could not be reversed by the specific antagonist flumazenil totalling up to 1.0 mg. Trauma related coma was diagnosed.

diagnostic exploration he had received a total of 80 mg of diazepam over the past 24 hours. Neurologic exploration was difficult as the patient now was in a deep comatose state. Computer tomography gave no clue in regard to intracerebral lesions, and since severe brain contusion was suspected, increments of the specific benzodiazepine antagonist 'flumazenil' were given in repetitive doses of 0.2 mg. The pathological power in the delta band, as visualized in the on-line EEG power analysis, could not be reversed, i.e. no fast frequencies could be evoked by the compound. Thus, the deep coma was referred to cerebral lesion from which the patient never recovered.

6.3 Symptoms of irreversible cessation of functions of the brain including the brain stem

A number of times monitoring of cortical activity is a prerequisite for diagnosing brain death. Apart from other symptoms (Table 3) the EEG ascertains the diagnosis displaying an isoelectrical line and no power spectra.

Table 3. Summary of those symptoms suggesting brain death.

1. Unconsciousness
2. No reaction to nociceptive stimuli
3. Loss of muscle tone
4. Maximally widened pupils with no reaction to light
5. Fixed divergence of both eyes
6. Respiratory arrest
7. Hypothermia or poicilothermic reaction
8. Loss of circulatory regulation
9. Spinal reflexes
10. Loss of brain-stem reflexes (gag, occulo-cephalic, and occulo-vestibularis reflex)
11. Loss of pupillary and corneal reflex
12. Negative apnoe test (an increase in CO_2 results in no respiratory movements)

Under the supposition that no muscle relaxant, benzodiazepine and/or barbiturate overhang is present, all of which may mimic some of the above-mentioned symptoms, and that an existing electrolyte disturbance is corrected, the EEG (with its power spectra) is a necessary tool in determining cessation of brain function. The neurophysiological evaluation thus comprises not only the effect of:

a) nociceptive, and
b) auditory stimuli

on EEG activity. In some centers additionally, *auditory evoked potentials (AEP)* with special reference to the early peaks (=brain stem auditory evoked potentials=BAEP) are mandatory for the determination of brain death which may or may not be followed by angiography of the carotid arteries to establish absence of cerebral blood flow (see also page 145).

6.4 Avoiding false interpretation of the EEG signal

Aside from these representative examples which underline the usefulness of compressed EEG analysis in the ICU, it is mandatory that during each period where power spectral analysis is performed the raw EEG signal should always be inspected beforehand. Otherwise, undetected artifacts would go into the process of EEG analysis resulting in false information.

One such 'false information' was derived in a 65 year old patient who had undergone prothetic implantation of the brachiocephalic truncus and postoperatively acquired mediastinitis and hypotension with renal insufficiency. After extensive neurological examination intravital brain stem dysfunction was diagnosed as there was a loss of brain stem related reflex activity.

In the computerized EEG analysis, a dominance of delta and theta waves was evident, with interpolated alpha activity suggesting viable cortical action (Fig. 44). It was only with the aid of the somatosensory evoked potential (SEP; see later) that a significant functional deficit could be diagnosed with a loss of the early peak, suggesting cortical and thalamo-cortical malfunction. As compressed cortical eletrical activity (CEA) only picks up potential changes from the outer cortical layers, a coma vigile was diagnosed. Thus, in certain situations it is necessary not only to rely solely on compressed EEG analysis, but simultaneously take into consideration

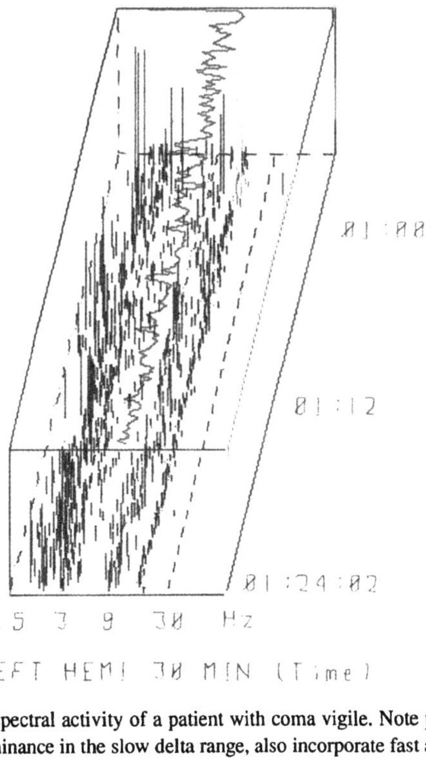

Fig. 44. Compressed spectral activity of a patient with coma vigile. Note power spectra besides depicting a pathological dominance in the slow delta range, also incorporate fast alpha activity which may lead to a misinterpretation of the patients status.

neurological tests and other neurophysiological ways of examining CNS function, such as the evoked potential (see page 84).

6.4.1 Electrical interference and the EEG signal

The need to inspect the raw EEG signal as often as possible is depicted in another example, where compressed power spectra were derived in a patient after successful resuscitation (Fig. 45). Aside from the pathological slow waves in the delta domain, regular 'jig-saw'-like fast alpha and beta activity is present. However, the missing power in the theta-domain raises suspicion for an interference with another electrical source. The search for the possible cause of such electrical interference may be troublesome as the ICU is contaminated by various electrical sources plugged to the socket, all of which act like small radiotransmitters.

Inspection of the raw EEG visualized that incoming signals were 'contaminated' with a sinoid-like slow wave which, when crossing the zero-line, was wrongly taken as delta activity. The cause of this 'contamination' was sweating of the patient, and once the electrodes were replaced by fresh ones (in such instances a different type of electrode should be used, preferably needle or cup electrodes),

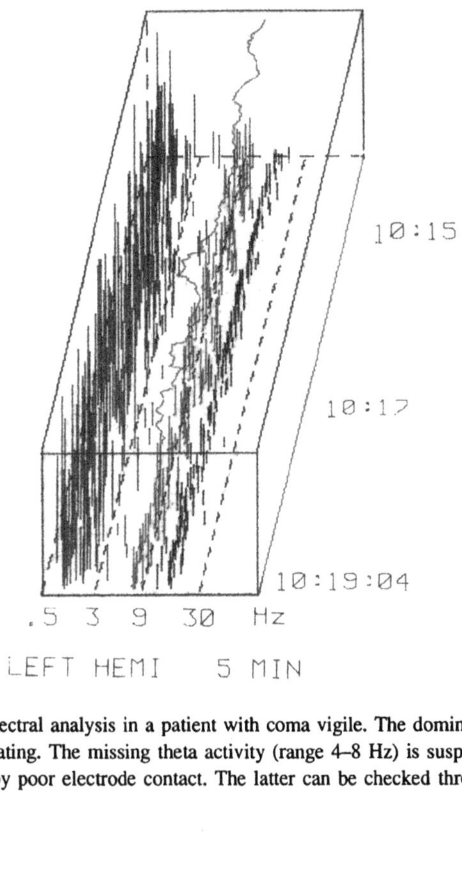

Fig. 45. Compressed spectral analysis in a patient with coma vigile. The dominant delta activity is an artifact induced by sweating. The missing theta activity (range 4–8 Hz) is suspicious for an electrical artifact being induced by poor electrode contact. The latter can be checked through the electrode impedance.

only fast activity was evident. Thus, coma vigile was diagnosed, as neurological tests and the missing slow wave components further supported the suspicion.

As mentioned before, cerebral monitoring in the intensive care as well as in the OR may become a problem when interference from other electrical appliances is present. These interferences evolve from:

a) an infusion and/or a perfusor pump,
b) a ventilator,
c) a dialysis machine,
d) a scan light,
e) a screen for direct reading of blood pressure and central venous pressure, and
f) the ECG.

6.4.2 Artifacts arising from the patient

But aside from these electrical interferences which are readily detectable as they are rhythmic in nature and induce a fluctuation of the base line, artifacts as they arise from the patient also have to be taken into account. An alcoholic with right-

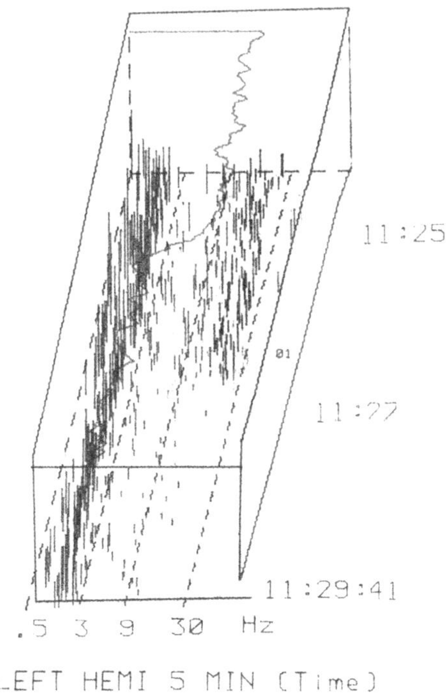

Fig. 46. Effect of tremor on EEG power spectra in a patient being sedated in the ICU. Injection of a muscle relaxant (event No 01) ceases power in the alpha and beta domain.

sided brain stem lesions (CT verified) who was being ventilated postoperatively after oesophageal resection, showed an increase of beta activity in the EEG. As suspected, the power in the beta range, which otherwise would only be found in an awake and not sedated patient, was due to muscle tremor. A small dose of a competitive muscle relaxant substantiated the suspicion, as shortly after the injection, power in the alpha and beta domains ceased completely, resulting in a dominance of pathological slow waves (Fig. 46).

6.5 EEG power spectra during sleep

Apart from the use of continuous, compressed and computerized electroencephalographic analysis (CEA) in the OR, the intensive care, as well as the emergency and the detoxification units, the sleep researchers may find this technique a useful adjunct for overnight monitoring in cases of insomnia. Due to a storage capacity of up to 10 hours and longer, the capability to store overnight EEG recording on disk makes these machines an attractive alternative to the usual strip chart recording.

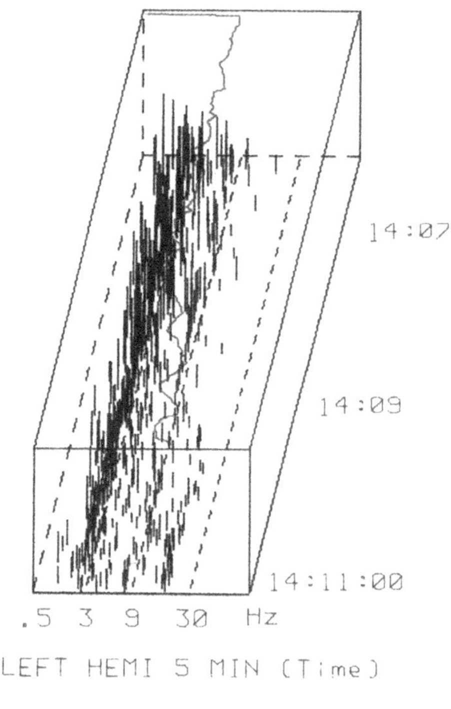

Fig. 47. Overnight sleep recording in a processed manner, depicting the change from slow wave sleep with high activity in the delta band to REM sleep. The latter is associated with activity in the fast beta domain and a reduction of power in the delta band.

A representational print-out of overnight sleep recording of the author shows the compressed and processed EEG data over a time axis of 5 minutes at 14.00 a.m. In Fig. 47 different stages of sleep can be visualized. At 14.09 a.m., fast frequency in the beta range is evident suggesting REM.

This change in EEG power pattern is accompanied by artifacts reflecting eye movements. Shortly thereafter, concomitantly with the apperance of fast activity, power in the slow delta domain drops suggesting a stage of REM (rapid eye movement) sleep which according to sleep researchers, primarily is associated with dream activity. This kind of processed recording has not been fully evaluated to its full potential in sleep research. Whether the compressed analysis in comparison to strip chart recording contains more information for the physician dealing with sleep disorders remains, however, to be proven.

7. Trouble shooting

This section is intended to aid the user of computerized EEG analysis in solving minor operational and technical problems which might arise in the course of processd EEG monitoring. Since in the majority of cases problems in recording are due to electrode malconnection and/or electrical interference, defaults which arise from the central processor of the internal circuit of the EEG device will not be dealt with.

1. Problem No 1 in all EEG recordings are artifacts which arise from the recording electrodes themselves. As mentioned earlier for any EEG channel that indicates an

impedance above 5 kOhm

the electrode on the skin has to be checked and possibly replaced. Before doing so, *again* the skin has to be cleaned thoroughly with alcohol from electroconductive gel and one has to make sure that the site of application is *completely dry.*

2. During the recording procedure a gradual increase in impedance is noted. This is a warning for an improperly functioning adhesive disc electrode which tends to come off. One can solve the problem by simply replacing the pregelled conductive electrode unit after thorough cleaning *and* drying the site of application.

3. Now and then there are cases where due to intense sweating of the patient any disc electrode will tend to come off (Fig. 48-A3). The EEG signal then is characterized by large fluctuations of the base line which in some instances is processed as EEG activity in the delta range. This is the only *exception* where in EEG recording the subdermal needle electrode is advocated. Additionally, the cup electrode which is fastened to the skin by a gluelike material (=collodion) does a similar job.

4. High static waves in the raw EEG signal are an example of interference of other electrical appliances. In such cases one should try to ground all other electrical units nearby as the electrical noise in the EEG is due to AC current flowing freely through the room (Fig. 48, B1–2).

5. Problems which arise during the recording period due to an uncooperative patient, or a patient who is in a hyperactive state will result in artifacts with excessive base-line wandering as demonstrated in Fig. 48, A1,2,4,6. In these instances the patient has to be sedated.

6. External interference by the body of the physician touching the patient and people walking across the room often tend to result in *static charges* (Fig. 48, C1–2) which will be picked up by the sensitive EEG preamplifier. One has to solve the problem by stopping everyone in moving about. Similar effects of static charging can be induced by movements of plexiglas near the preamplifier (Fig. 48, C3).

7. Electrode cable connection and breakage tends to induce fluctuations which are either identified as artifacts or processed as slow waves in the delta range. Thus, if there is any doubt about the results as they are processed from the raw EEG, one should check the raw EEG waves as they are fed into the computer. Such a procedure usually is very simple by activating a touch key indicating 'raw EEG signal', thus identifying the cause of malfunction (Fig. 48, D1–4).

8. Excessive artifacts of the raw EEG signal may be due to a number of reasons:
   a) accidently jarred leads of the preamplifier,
   b) gross muscle movements,
   c) electrocautery,
   d) 50–60 Hz power line interference,
   e) electrodes not properly prepped or gelled, and
   f) faulty electrode leads.

Prior to all processing, the above problems have to be eliminated by making certain that electrodes are *prepped*, that electrode leads *do not run across a power line*, that faulty electrodes are replaced. 50–60 Hz electrical noise can easily be identified on the raw EEG signal. One has to localize the source and *switch off* the electrical machine.

Electrocautery will *result in absolutely no proper recording* with any device. In such a case, the electrical current running inside the organism is picked up by the electrodes and the signal fed into the preamplifier. For optimal recording one has *to wait* till electrocautery is finished. It should be borne in mind that improper grounding *of the electrocautery* unit may result in local skin *irritation at the site of the EEG electrodes!*

However, sufficient electrode gel between the electrode disc and the skin will prevent this from happening.

9. If no EEG signal is displayed on the EEG monitoring screen although the unit is turned on, one should check if the electrode cables are connected to the preamplifier and the preamplifier socket is connected correctly to the main unit. In rare cases a faulty preamplifier or a disrupted lead is the cause of such a problem.

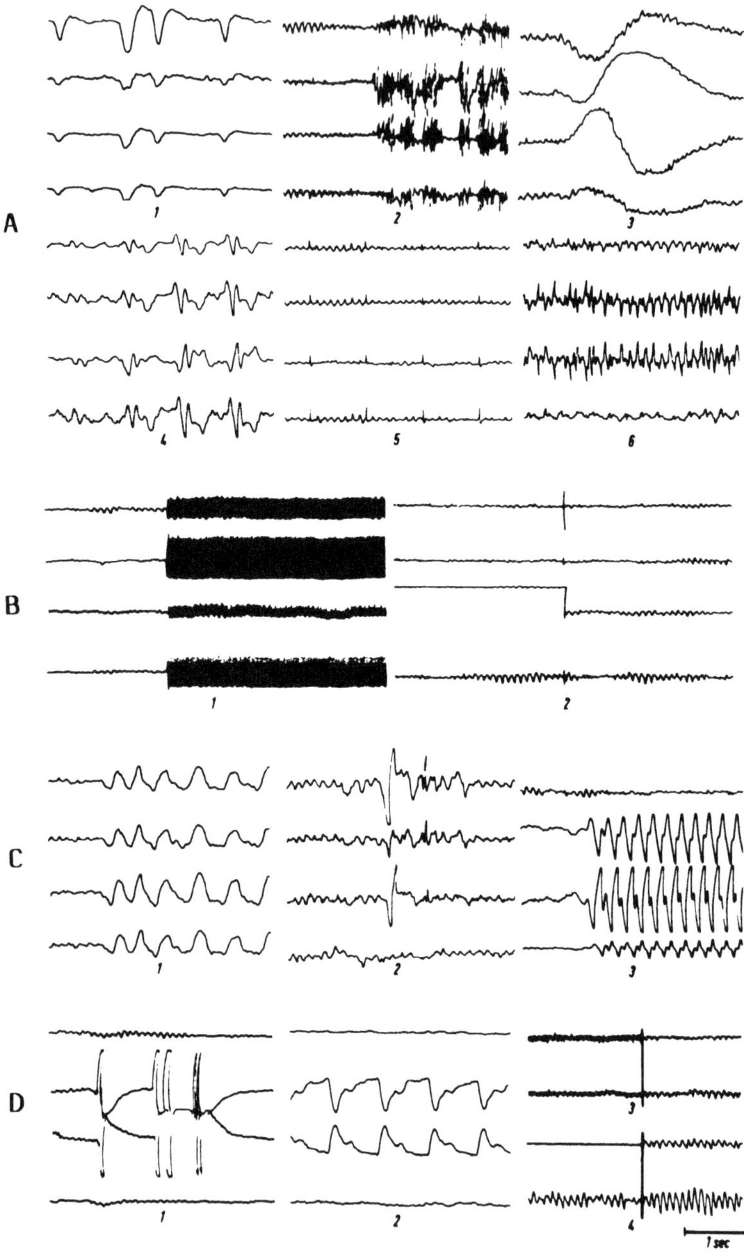

Fig. 48. Summary of the possible artifacts which interfere with the original EEG waves coming from the patient.

To Fig. 48. Summary of the possible artifacts which interfere with the original EEG waves coming from the patient.

A. Biological artifacts

1. Eye lid movement
2. Muscle activity during swallowing
3. Sweating, movements of arms and legs
4. ECG cables running across EEG leads thus picking up myocardial activity from the heart
5. Muscle tremor

B. Interference with other electrical appliances

1. AC current due to improper grounding

C. External interference from other people in the room

1. Movements in the room where EEG is taken
2. Paces in the room where EEG is taken
3. Plexiglas is being moved on table near the preamplifier

D. Electrodes and connections

1. Artifact due to cable movement
2. Unproperly fixed reference electrode during bipolar recording
3. High impedance
4. Faulty insulation of electrode

---

8. Systems currently available for processed EEG recording

8.1 Introduction

As the OR or the cubicle in the ICU where the patient is to be monitored usually is filled with other necessary instrumentation for monitoring and anesthesia equipment (i.e. respirator, anesthesia machine, ECG recording device, heart-lung machine, cell saver, etc.), not much space will be left for an additional monitor to record and process EEG activity. Thus, one of the major prerequisites for the practical use of processed EEG information during anesthesia and post operative care are:

1. *Compactness* and *mobility* of the unit which has to fit into a condensed room, preferably on wheels to be moved around easily.
2. Ease of operation and ease of interpretation of the compressed EEG data even by a *non-specialist*, thus avoiding long-lasting introduction in handling the unit and requiring trained specialists.
3. Flexibility and a number of options in processing the raw EEG signal.
4. Incorporating the option of additionally analyzing somatosensory-evoked potentials (SEPs) or another form of evoked potential monitoring (see later).
5. Storage capacity of compressed EEG data on disk or tape from which data can be retrieved easily for later review.
6. Continuous automatic impedance monitoring to affirm sufficient electrode contact.
7. Sufficient filtering to eliminate 50–60 Hz power line interferences.
8. Fast-copy capabilities with a built-in printer.
9. Expanded memory capacity for long-term recording of up to 8 hours and more.
10. Simultaneous display of both hemispheres with trending displays.
11. Data display with sufficient contrast on the viewing screen either in a numerical and/or graphic mode, according to the users demand.

12. Easy switching between raw EEG waves, compressed analysis and evoked potential operation.
13. Artifact detection and rejection fast enough to ensure reliable EEG interpretation.
14. RS-232 coupler interface for expanding to raw EEG and preprocessed data analysis
15. Comment option in the final print-out of the processed data.
16. Low cost in maintenance and sufficient as well as fast service by the local distributor.

Among the units available on the market, some meet most of the above mentioned prerequisites for processed EEG monitoring, while others tend to lack some of the essentials which are necessary for EEG processing in the OR and/or the ICU.

The following is an updating of those recording units which are claimed to be useful in the clinical sense, meeting the latest standard in computer philosophy. However, due to the rapid development in computer and microchip technology, other devices are bound to be developed which contain numerous other options in a more condensed and more applicable form. Up to this moment all the units in the next chapter have been in use in the OR and/or the intensive care where their capabilities have been tested on practical grounds. Before going into details it should be mentioned that numerous low-cost systems are not included as they do not meet the above prerequisites which make EEG data management easier in the daily routine work.

8.2 Description of units used for EEG power spectral analysis in the OR and the ICU

The Neurotrac^{TM} (Interspec Medical, Wisconsin, USA)

This unit is a compact system which offers a large spectrum of possibilities:

1. Compressed Spectral Analysis (CSA) displaying EEG frequency and amplitude data in 'hill and valley' format as trends (Fig. 49).
2. Spectral Edge Frequency (SEF), a monoparametric descriptor of power spectral analysis, is useful as a guideline of the activity in each epoch. It is a marker in the outer edge of the CSA display providing easy recognition of changes in the higher frequency level in relation to power. It usually incorporates 95% of all EEG power. However, a change in the spectral edge, selecting any number from 1–100%, can be ordered through the keyboard input.
3. Spectral histograms; they display the power in half hertz resolution in the format of a bar graph. The histogram is continuously updated at a preselected rate (Fig. 49).
4. A power band presents quantitative power values in any of the four user-selectable ranges (usually delta = 1–4 Hz, theta = 4–8 Hz, alpha = 8–13 Hz, and beta = 13–30 Hz).

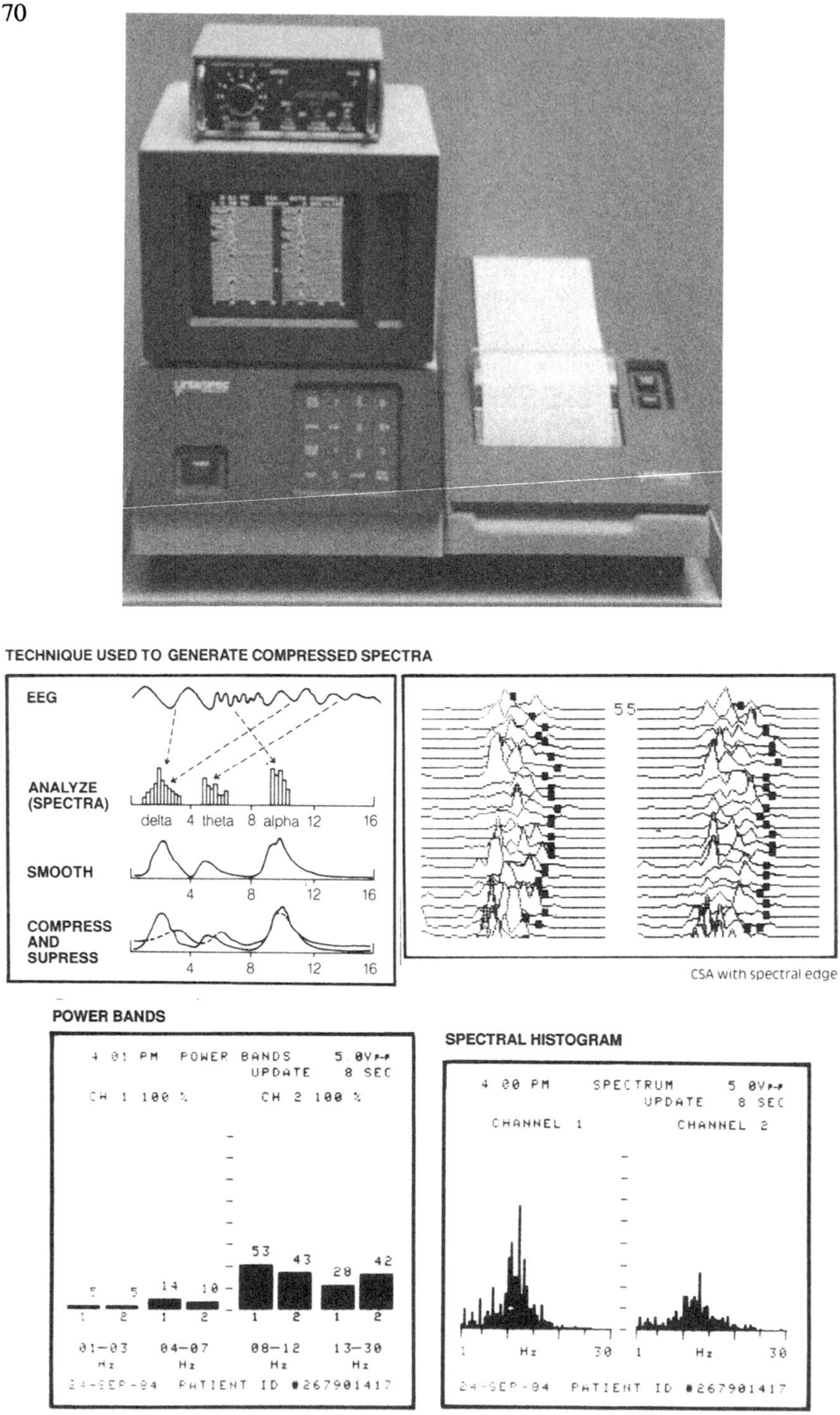

Fig. 49. Front view of the Neurotrac™ Cerebral Monitor which uses the FFT algorithm for EEG analysis with examples of CSA, spectral histograms and power bands.

5. Somatosensory evoked potentials (SEP), an optional neurophysiological measurement system to detect changes in the functional integrity of the sensory nervous system during continuous monitoring in the OR or the ICU. (For further details see Chapter 9.)
6. Electrode switching box permitting rapid selection of different electrode montages in the EEG and/or the SEP mode.
7. A separate fast black and white printer using thermo-sensitive paper.

Advantages:
1. Compactness with small dimension, which permits easy application in the OR and the ICU
2. Multiple display function that fit the average need
3. Simplified operation solving the difficulties of conventional EEG equipment.

Disadvantages:
1. Capacity of data storage only up to two hours.
2. Thermo printer paper has insufficient contrast and fades in time
3. Sampling rate of 128 Hz/second sometimes may result in insufficient resolution of the EEG
4. Monochromatic display on screen is tiring during long-term viewing
5. Relevant power in the CSA mode is hidden behind peaks of preselected activity of power spectra.
6. Spectral histograms are of little practical use
7. To recall quantified EEG power spectra of the past, one has to scroll through all previous data
8. The storage of data on floppy disk is only possible through an external computer.

Comment:
If no specific tasks such as long-term monitoring and research has to be done with the unit, it represents a useful item in CNS monitoring as it can be expanded to evoke potential monitoring.

The Lifescan™ Monitor (Neurometrics, San Diego, California, USA)

This unit also is characterized by its compactness and mobility which is underlined by the fact that it comes on wheels. Recording of EEG is possible even during transportation of a patient. The outstanding feature of this unit is the real-time wave-by-wave analysis (Aperiodic Analysis) which, in contrast to the usual FFT, purportedly results in faster response and a higher resolution of EEG waves.

Following facts differ from many other devices:
1. A 30 cm diagonal full color display with a three-dimensional picture of power and frequency changes in vector display. The different power bands

are coded by different colors thus making long term and long distant viewing easier (Fig. 50).

2. A remote control colored and numbered function key with infrared communication to the computer permits easy change of formats from a distance.
3. A built-in color printer for hard copy.
4. An expanded memory of approximately 5 hours in the 5 minute mode, 10 hours in the 10 minute mode, and 24 hours in the 30 minutes mode, plus 6 user-selectable snapshots which can be stored at any time.
5. An option for the somatosensory evoked potential mode to assess neurological function.
6. A memory protector of 20 minutes during which EEG data can be stored and analysed, such as during transportation, without the use of a power outlet.
7. A RS-232 interface to connect to a Kaypro™ 2000 PC (or any other comparable computer with a math processor except AT's and XT's) with special software as an option for the quantification of EEG data. A useful item especially for research purposes.
8. Storage of snapshots on the computer disk are possible for later retrieval and visualisation on the screen.
9. A special software package permits individual selection of methods for further procession of raw EEG data.

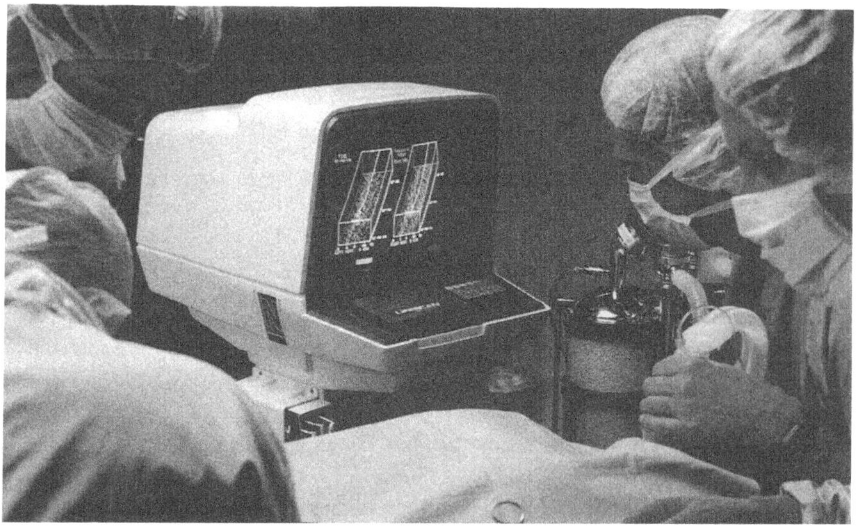

Fig. 50. Front view of the Lifescan™ EEG monitor. The 'glass boxes' on the screen reflect the two hemispheres where data is being displayed. The unit comes on wheels permitting mobility.

Advantages:

1. Aside from the preported superiority to the 'aperiodic analysis' in comparison to FFT, a claim which still has to be proven under clinical conditions, display of processed data on the Lifescan screen is by far the best, especially in view of the separate colors to delineate each frequency band.

2. The user selectable 'activity' edge frequency which moves along as a white line as time passes, is a useful monoparametric parameter, e.g., in evaluating the depth of anesthesia in most anesthetic techniques being used.
3. The remote control which permits a rapid change of data display by simple touching of softkeys.
4. The expanded memory permits overnight recording with the possibility of retrieval of all processed data in the morning for inspection.
5. The high sampling rate results in an optimal resolution of EEG waves, a fact which becomes more evident when evoked potentials are derived.

Disadvantages:

1. One major disadvantage is the slow speed of the built-in printer which also tends to be very sensitive to improper handling. A recently improved system incorporates a fast, robust thermal printer with enhanced print capabilities and a better resolution.
2. The need for an additional labtop computer (such as the Kaypro™) for the quantification of data such as in research.
3. The electrode impedance, although checked continuously and indicated by a colored band, cannot be checked separately for each electrode.
4. Although transportation within the house is very easy as the unit is installed on wheels, transportation to outside locations is difficult as the stand with the wheels cannot be separated from the main unit.
5. The top main unit is relatively heavy; wheeling over a small shoulder may result in imbalance and tilting of the whole device.

The EEG Trend Monitor (Nihon Kohden, Japan)

This is a two-channel analyser, processing the raw signal into a compressed spectral array (CSA). The system's main feature is that it provides a single-glance comprehensive format in frequency and amplitude of the EEG waves over time in the form of a CSA display and/or a (freezable) thermal printing. Fast Fourier Transform technique is applied to create a frequency spectral wave form of each second for compressed spectral array. Frequency spectrals are averaged every 10 seconds to 120 seconds maximum and printed as an array. The features include:

1. Simple operation via touch keys which are grouped by function on separate areas on the control panel.
2. Consecutive recording in any of the following three modes is possible:
   a) power spectral CSA,
   b) amplitude spectral CSA (this mode merits the correspondence of the amplitude of the raw EEG waves with the peak of the CSA),
   c) and percent frequency band where the total energy of spectral power (= 100%) is assigned to each frequency band (0–32 Hz) in proportion to the energy in each frequency component.

3. Optional cassette data recorder which permits recording and data reproduction with a maximum storage capacity between 5–60 hours depending on the analysis time.
4. Display of the original EEG wave on the monitor screen with a 25 or a 30 mm/second sweep speed.

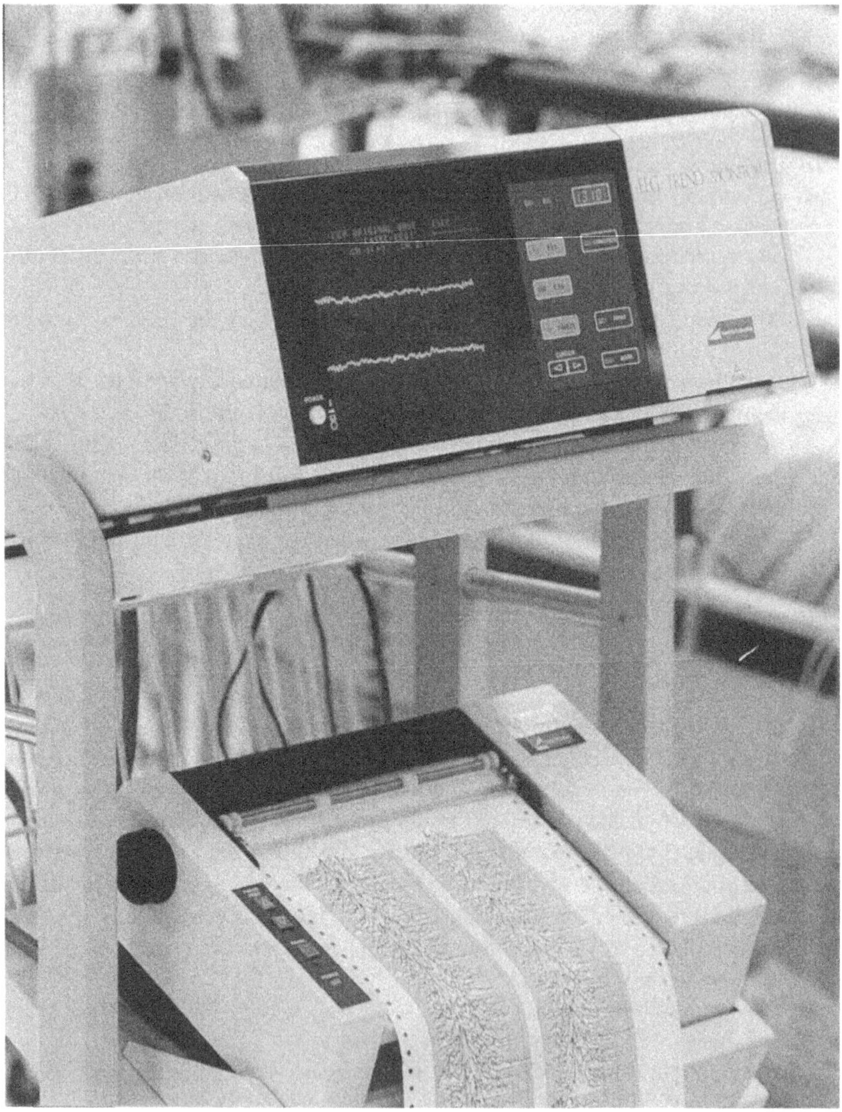

Fig. 51. Front view of the EEG Trend Monitor from Nihon Khoden with the print, both being housed in a cart for the ease of mobility.

5. Artifact-affected averaging with a magnitude over a certain predetermined limit automatically is rejected
6. Input/output terminals for system expansion permit to accept raw EEG signals from a conventional electroencephalograph, tape reader, or radio telemetry instrument for CSA analysis. The output terminal has an analog output and a GP-IBM connection for communication with a PC.

The unit can only be used for power spectral analysis of the EEG in the CSA mode. Thus, certain limitations have to be taken into consideration. However when it comes to ease of operation and especially when it comes down to the price of the 'trend monitor', both of them are in an acceptable range. The unit is especially designed for those users who do not want to go too deep into cerebral monitoring. There is no possibility to expand the system into the evoked potential mode. The percent frequency band may give some idea about the shift of power in the various frequency ranges over time, however there is no way to quantify these changes for research purposes. Mobility is guaranteed as the main unit and the separate high-speed plotter can be put into an exclusively designed cart for convenience of operation in the OR or bedside monitoring (Fig. 51).

The OTE Biomedica, Berg-Fourier Analyser (Farmitalia Carlo Erba, Italy)

Similar to the above-mentioned systems, it operates with two channels using the CSA mode (8-second epochs with 7 seconds overlap by FFT) and incorporates the option of quantification of power in the various frequency bands (μV^2). The user-selectable frequency bands are preset together with a time interval, computed and later printed on thermal paper. CSA power spectra can be viewed on a scope monitor (freezable) and plotted on paper with a heated pen using the built-in printer with four different speeds. A dual-channel cassette EEG recorder with patient input box permits remote recording and later analysis. A trolley designed for easy transport of the otherwise bulky system with a work bench and two storage drawers provides additional mobility and versatility. A special addition is the feature of artifact reduction arising from electrocautery. The latter can be suppressed by a separate foot switch control which is coupled with a remote reject input connection.

In a more modern version of the Berg-Fourier Analyser the quantification unit is replaced by the Post Elaboration Unit (PEU) with a thermal printer on a control keyboard (Fig. 52). The new design incorporates a number of advanced technology:

1. Power spectra are expanded to 356 second epochs.
2. A complete spectral measurement may be performed in absolute values.
3. Calculation of mean power spectra or epochs is selectable in one sec steps from 8–356 seconds.
4. Presentation of single spectral values with indication of amplitude ($\mu V^2/Hz$) and frequency (Hz) is possible.
5. Value trends resulting from the power of 12 selectable bands can be presented. Each value is expressed as a percent of the total power or of the total power of one of the calculated bands as selected by the operator.

6. Extended display capabilities of trends of a single band with measurement of the reference power in percent value of each epoch.
7. Trends can be obtained in real time corresponding to a 1 second epoch.
8. Mean spectra, expanded spectra and individual trends can be stored in 8 memory banks. As the unit functions are programmed through a keyboard the post elaboration unit is meant especially for neurological and pharmaceutical research adding little to the day-today work of the clinician.

Similar to all other units using FFT as the key algorithm for processing and displaying power spectra, some of the information may get lost when power is hidden

Fig. 52. Front view of the Berg-Fourier Analyzer (OTE Biomedica) and its recent extension the Post Elaboration Unit (PEU).

behind a previously high peak. The latter disadvantage, however, may be compensated by a special function in the Post Elaboration Unit (PEU).

This relatively bulky EEG analyzer is strictly meant for EEG measurments; it is not possible to extend the unit into an evoked potential system. The price is attractive, although the system with a sampling rate of 128 Hz/channel present some limitations in regard to resolution of the EEG wave.

The Biologic Traveler™ (System Corp., Illinois, USA)

The latest, a newcomer on the European scene, comes as a single-channel modality. It can however be expanded to a multi-modality four-channel system. Surprisingly this system is a small, portable and lightweight piece of equipment (Fig. 53). In addition to EEG analysis, the Traveler is wholly IBM-PC compatible which enables the user to operate a multitude of software programs in conjunction with this EEG analyzer through the detachable keyboard. In addition to the routine CSA mode, the system incorporates evoked potential testing, electronystagmogram (ENG), electroocculogram (EOG), bone conduction, ECOG, and 40 Hz ERP (see Chapter 15) capabilities. The built-in AC/DC amplifier makes the system capable of EMG and EOG testing, eliminating the need for separate recorders. The Traveler can also be expanded to include 'Topographic Brain Mapping' (see Chapter 19), a recent development in brain wave imaging, and the early detection of malfunction and/or malperfusion.

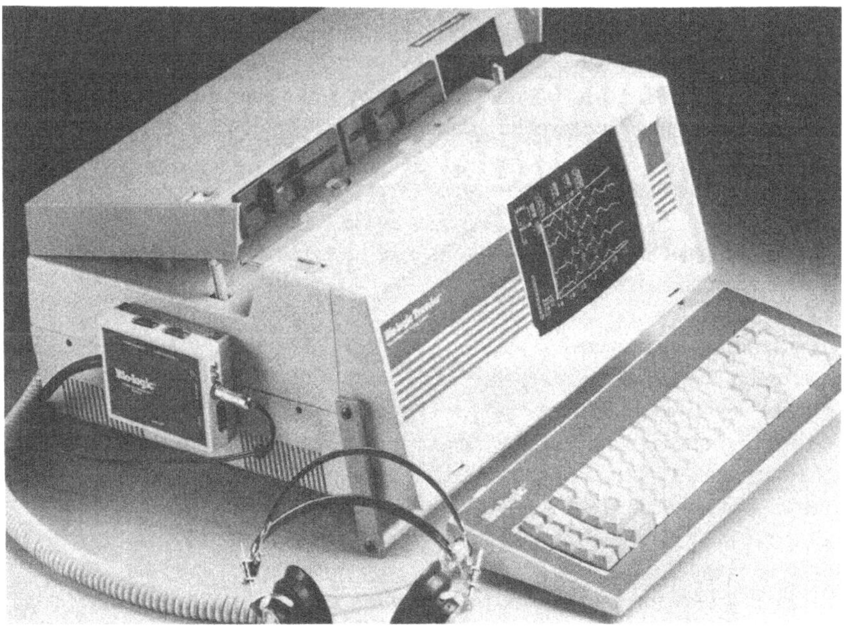

Fig. 53. The Biologic Traveler™ which is portable and can be used for various electrophysiological measurements.

The dual-channel 16-bit microprocessor provides remarkable capabilities (320 kbyte of dynamic RAM expandable to 640 kbytes). Two floppy disk drives, storage capacity of 2 × 360 K, which permits waveforms and patient data to be stored along with the EEG. All vital data including patient records and comments are stored automatically and permanently. With the use of the printer buffer, data can be printed or plotted using a separate plotter.

The display is an amber phosphor screen which may be connected to a color video screen. Waveform display of averaged signals with simultaneous raw EEG data is possible.

Due to its expandibility (17 different system operations) and a number of options available to tailor the system to the individual demand, sophistication is a reality. However, due to these various capabilities the clinician not so much in rapport with computer technology may find it difficult to manoeuver his way through the variety of features being offered. Thus, the Traveler is meant for someone who wants to extend his capabilities in measuring other neurophysiological parameters, not just concentrate on EEG analysis. This is something the computer-orientated clinician has long been looking for, who likes to dwell upon all forms of functional monitoring of the CNS.

The two channel EEG monitor can display raw EEG waves, spectral edge frequency, time plots, the average intensity in each of the 6 wavebands and numerical analysis simultaneously. Up to 16 channels of processed EEG may be displayed on 12 screen display formats. The screen can be split to display combinations of trend analysis, numeric data or CSA graphics. Impedance checking is easily done with the press of a key selecting *one* electrode on a real time basis. For data storage one can record directly on a computer disk drive, with mass storage for 24 hours monitoring. Simultaneously for the record, CSA from the color monitor can be printed on-line with a separate color graphics printer.

The Dr. Richard Weiss Neuromonitor™ (West Berlin, F.R.G.)

This is a computer-based brain monitor which incorporates other forms for vital cerebral monitoring such as:

a) intracranial pressure,
b) arterial blood pressure,
c) cerebral perfusion pressure, and
d) transcutaneous and cerebral pO_2/pCO_2 measurements.

The two EEG signals which can be viewed on a screen will be processed via FFT into power using 4 second epochs, thus yielding CSA which is displayed and printed in a chrono-spectrum. Additionally:

a) the relative band power in relation to the sum power of the 4 frequency bands is computed,
b) the absolute sum power in the frequency range 0–30 Hz is displayed,
c) the mean frequency, and
d) the dominant frequency are calculated.

The advantage of this system is the simultaneous measurement of EEG and relevant pressure dynamics as well as their interrelated effects. Thus, interactions of intracranial pressure changes and the EEG can be correlated with each other. The unit is especially designed for the monitoring of head trauma patients in an easily operated joint function providing data storage on floppy disk. This includes qualitative-numerical presentation of the most important EEG parameters and the interfacing with a background computer. An expansion of the system incorporating evoked potential testing is under development. Additionally, a sleep polygraph recording system will eventually be added giving more information in regard to cerebral monitoring in special research situations. Although not too light-weight, the unit still has a dimension that allows it to be placed on a trolley for mobility.

Fig. 54. Front view of the Dr. Weiss Neuromonitor™, a system which can be used for extended monitoring of head trauma patients.

The Cerebro Trac™ 2500 (SRD Medical,UK)

Next in line incorporating the more recent developments in cerebral monitoring is the Cerebro Trac which combines a number of features meant for the daily use:

1. Real-time FFT spectral analysis of two channels in color
2. Density Spectral Array (DSA) permitting fast recording at a glance reading. Perportedly an improvement over the older CSA display method, it is said to have an advantage as total data presentation in increased legibility is now possible.

3. Frontalis muscle EMG as an indicator for anesthetic depth and of neuromuscular blockade. It measures trend alongside the EEG.
4. Spectral Edge Frequency which, as already mentioned, indicates the highest domain frequency in the signal and is a useful correlate of anesthetic depth.
5. Numerical as well as long-term trended display formats.
6. An optional ink-jet printer (from HP) providing permanent storage of patient's records.
7. Simultaneous display of raw and processed EEG data

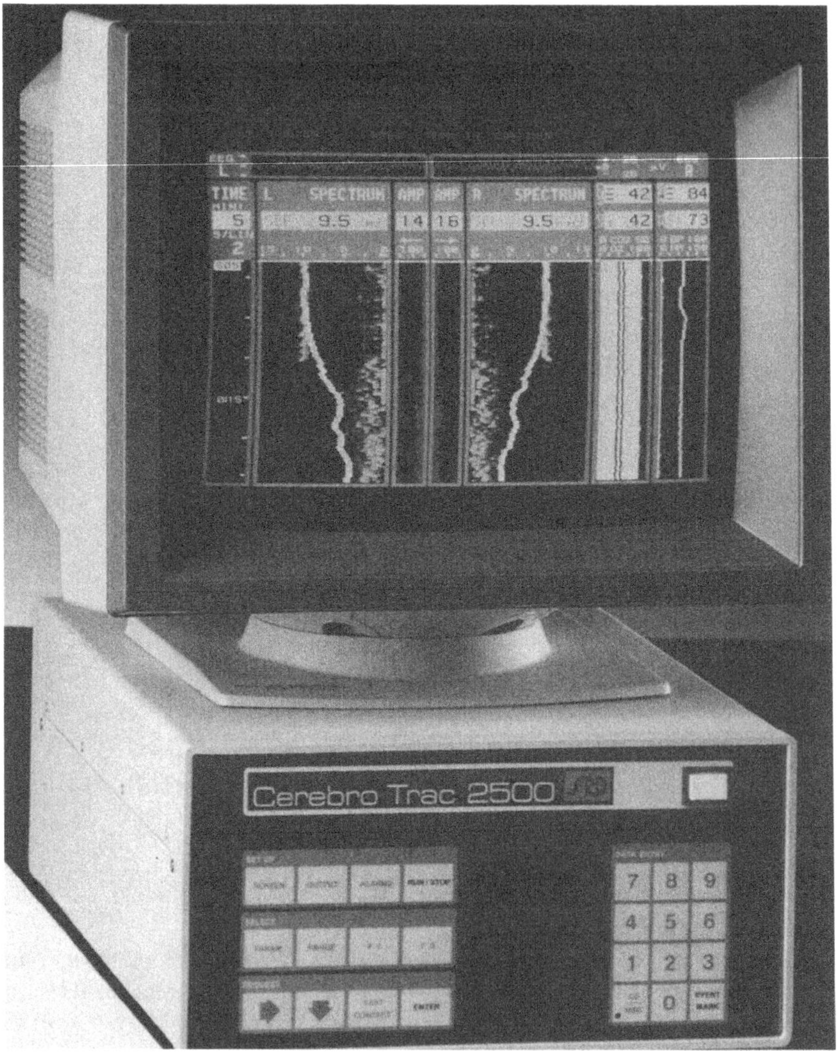

Fig. 55. The stationary Cerebro Trac^{TM} 2500 with various parameters all on one screen. Operation is made simple through touch keys.

8. An interface with existing monitors to trend physiological parameters such as:
   a) heart rate,
   b) blood pressure, and
   c) temperature,
   which permits comparison of processed EEG data with a synchronized trend of other vital signs.
9. Alarm limits to alert to sudden changes in cerebral activity or electrode fall-outs and an auto-checking of electrode impedance.
10. An RS-232 serial port which permits mass storage of the entire data streams on-line on floppy disk (IBM-PC or compatible).

This system is as a pure monitor of cerebral activity to be combined with other vital parameters (Fig. 55). The EEG analysis is much orientated in regard to distinguish between amplitude and frequency changes comparing both hemispheres with each other. Not meant for portability, it is a system for stationary recording and easy to operate. There is no possibility to incorporate the evoked potential mode.

The Nomad™ Portable Neurodiagnostic System (Tracor Northern, Wisconsin, USA)

This is a four-channel recording system for processing EEG. The quantification mode includes:

1. Spectral Edge Frequency.
2. Color Density Spectral Array (CDSA) permitting three parameters (time, frequency and magnitude of activity) to be individually displayed.
3. Magnitude, dominant peak and frequency ratio, including a % calculation of power in the frequency bands, is computed automatically.
4. Processed data may be stored on floppy and/or hard disk.
5. A hardcopy output for patient record using a HP printer is optional.
6. The system can be configured in such a way as to include independent somatosensory stimulation permitting *simultaneous* acquisition of responses from up to four different stimulation sites.
7. As needs grow, other features can be added, including real time EMG (electromyogram), EOG (electroocculogram), ERG (electroretinogram), and CNG (cochlear nystagmogram).

The system by itself thus covers practically the whole range of a diagnostic set necessary in neurology (Fig. 56a). It demands an experienced user to take advantage of all the features it offers besides EEG monitoring. For a department which asks for a unit to cover all aspects of neurological monitoring, this is a remarkable piece of equipment despite the fact that portability may become a problem. The latter is solved by a cart upon which the device is mounted. A potential drawback is the fact that although customer support is provided by a trained staff in the US, distributors on the European market are sparse.

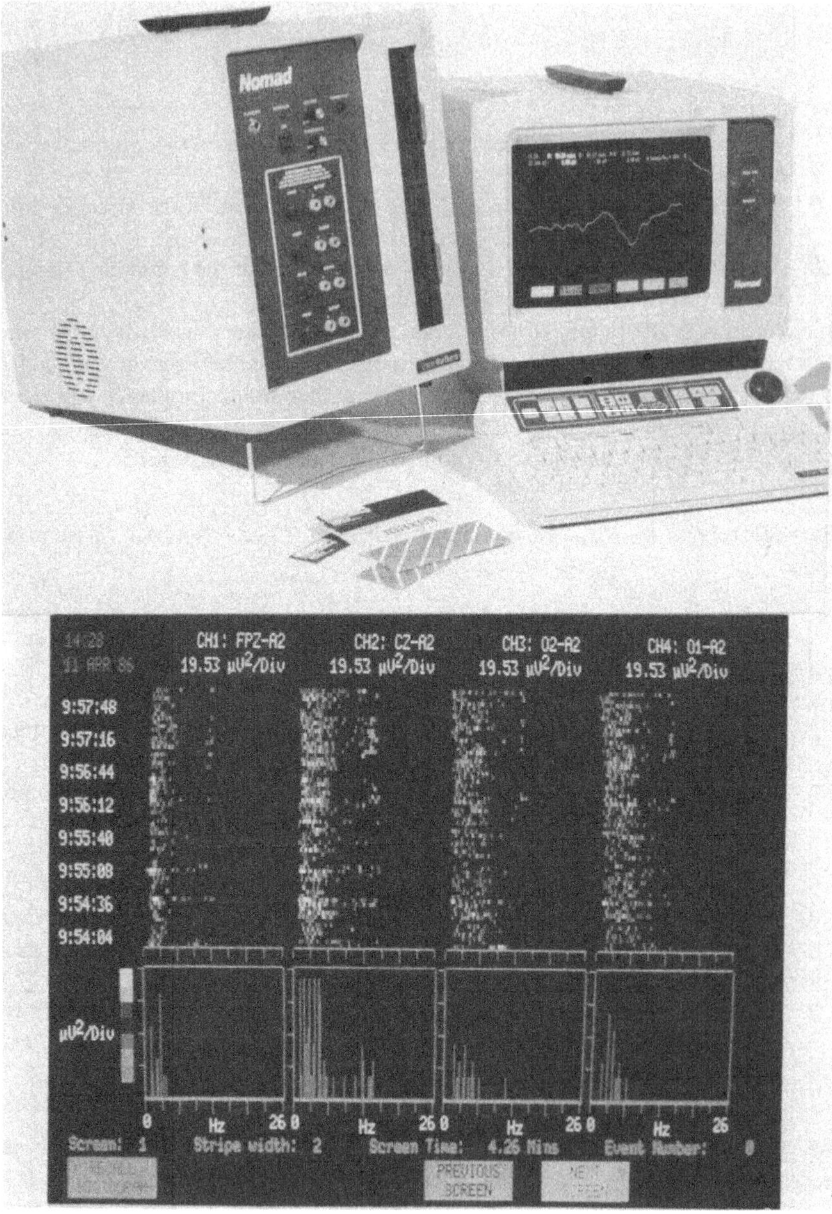

Fig. 56a. The Nomad™ with its main three parts can be mounted on a chart to ease transportation. The lower part shows a typical display of the CDSA denoting three parameters, time, frequency, and magnitude of power (µV²/Div) in the various frequency bands.

Recently Tracor Northern introduced the Guardian™ Brain Activity Monitor, tailored to the anesthesiologist's need, which monitors adequacy of cerebral perfusion, depth of anesthesia and is also suitable for long term monitoring using up to four channels.

The following mode of processing and quantification are offered with FFT signal processing:

1. DSA with a resolution of 0.25 Hz or 0.5 Hz per frequency bin
2. Mass storage on a 20 MB hard disk
3. Trend charts calculating in real time
   - total magnitude of each channel
   - relative magnitude in any frequency band
   - percentage of total magnitude of any specified frequency band
   - dominant frequency
   - ratio of frequency bands, and
   - epoch averaging of trend

A HP Thinkjet™ printer for hard copy, an external streamer to increase storage capacity and a mobile cart are optional.

Fig. 56b. The light weight and compact Guardian™ Brain Activity Monitor with touch screen user interface and up to 22 hours DSA data storage.

9. Sensory Evoked Potentials

9.1 What they are and what they offer

As previously stated, the EEG presents the electrical activity from only a small portion of the cortex. Furthermore, the varying changes of different anesthetic agents and the different depths of anesthesia can be monitored. Aside from using the spectral edge, more sensitive methods can be used to fully understand the way in which anesthetics act on the brain, and more so, *how function is impaired.*

One such method is that of *evoked potential* measurement (Fig. 57).

The principle of the method is to apply a stimulus and then to record the electrical change of the signal along its specific pathway up to the corresponding region of the brain. The stimulus may be of either visual, auditory or somatic nature.

In case of the latter the stimulus is a small electrical potential applied to the skin of the hand. Thereafter, the stimulus travels along the specific nervous pathways inducing (= generating) potential activation at various sites.

The generation of potential changes at various sites along the pathway is an index for the *integrity* of the nerve. Thus, the evoked potential can be considered a neurophysiological response (usually of the cortex) to impulses originating from some externally stimulated sensory nerve. They provide a physiological measure of the *functional integrity* of the sensory nerve pathway which can be used as a clinical diagnostic tool as well as for intraoperative monitoring. The evoked potential usually is recorded from the specific cortical area corresponding to the stimulus input. For instance the stimulation of sensory nerves will result in impulses travelling along the peripheral nerve, the spinal cord and finally ending at the contralateral, post-Rolandic sensory cortex (Fig. 58).

An evoked potential differs from the EEG mainly in 2 ways:

Fig. 57. With the introduction of evoked potentials, a number of clinicians face a completely new methodology.

1. The EEG is a random, continuous signal which arises from the ongoing activity of the *outer* layers of the cortex. An evoked potential is the brain's response to a repetitive stimulus along a specific nerve pathway.
2. EEG signals range from 10–200 μVolt. Evoked potentials are smaller in amplitude (1–5 μVolt) requiring precise electrode positioning and special techniques (signal averaging) to extract the specific response from the underlying EEG 'noise'.

The technique of signal averaging, as originally described by Dawson in 1954 [69], has been further developed in computer processing. The technique is now used by applying a stimulus repeatedly – but preferably at randomized intervals – and to record the evoked response over the corresponding area of the brain, averaging out mathematically the change over the number of stimuli.

9.2 Rationale for the use of EPs in the OR and the ICU

Evoked potentials (EPs) serve the following major purposes:

1. Monitoring of the functional integrity of neural structures that may be at risk during, for instance, ECC (extracorporeal circulation) or endarterectomy indicating cerebral hypoxia.

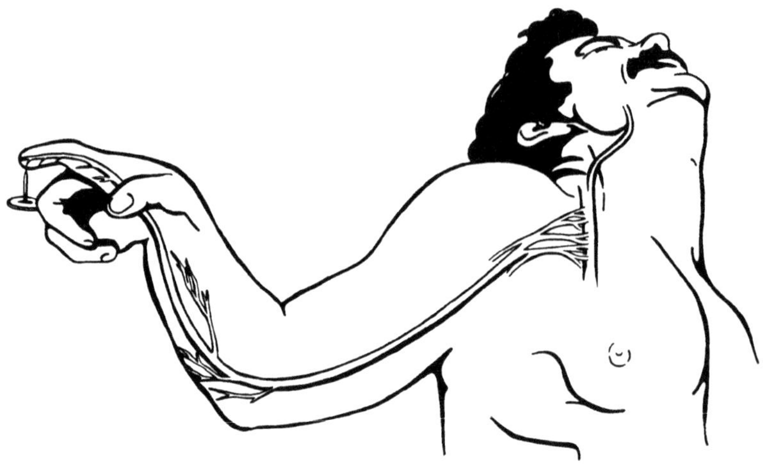

Fig. 58. Principle of the pathway of the somatosensory stimulus which can be recorded along the sensory tract.

2. Monitoring of the effects of anesthetic agents and other centrally active drugs which, besides the cortex, affect deeper neuronal structures.
3. Orthopedic cases where the spinal cord is at risk such as Harrington rod insertion and removal.
4. Clamping of the abominal aortic artery during aneurysmectomy resulting in a potential damage of the lower parts of the spinal cord.
5. Clipping of an intracerebral aneurysm which may be impeding blood flow to vital cerebral textures.
6. An indicator of cerebral hypoxia when the blood pressure is deliberately lowered.
7. Operation on peripheral nerves and nerve roots to identify early trauma.
8. Monitoring the cerebral function during controlled hypothermia when the EEG becomes flat.
9. Monitoring of the pathophysiological conditions after severe head trauma and the effects of therapy.
10. An intraoperative warning device of unsuspected awareness during light anesthesia when movement is abolished by muscle relaxants and cardiovascular responses are modified by vasoactive drugs.

9.3 The classification of EPs

If the auditory nerve is stimulated by 'clicks' from headphones, it is called the auditory evoked potential (AEP). The early part of the AEP waveform (less than 10 msec) is called the Brainstem Auditory Evoked Potential (BAEP) since it reflects the passing of the impulse through the brainstem.

If a nerve on the arm or the leg is stimulated by a small electrical current applied to the overlying skin, it is called the Somatosensory Evoked Potential (SEP).

If, however, the retina is stimulated by means of flicker light or a sudden change in a checkboard pattern, the evoked potential thus recorded over the corresponding cortical area is called the Visual Evoked Potential (VEP).

Evoked potentials are used both as a diagnostic tool and as a monitoring technique. As diagnostic tests, EPs are useful to evaluate neurologic disorders such as:

a) multiple sclerosis,
b) acoustic nerve tumors, and
c) optic neuritis.

As a monitoring technique, EPs are used during all surgical procedures which might compromise part of the brain or the spinal cord.

10. The principle of Somatosensory Evoked Potential Monitoring

There are distinct differences between the recording of evoked potentials and the recording of an EEG for spectral analysis:

a) The electrode *must* be placed on specific locations on the scalp necessitating the use of conventional cup or needle electrodes and *not* ECG pad electrodes.
b) The evoked potential waveform is produced by repeatedly averaging segments of the EEG *following* a sensory stimulus. Thus, a specific and repetitive stimulus must be provided and there will be a lag of up to several minutes before the results of the test are known.

10.1 Electrode types

In recording SEPs there are two sets of electrodes used. One is for detecting and for recording the impulse, while the other set is for providing the electrical stimulus to the nerve.

For stimulation, regular self-adhesive ECG electrodes can be used or a dual pre-gelled and self-adhering PNS (peripheral nervous surface) electrode is recommended (Fig. 59; NDM Corp, Dayton, Ohio, USA).

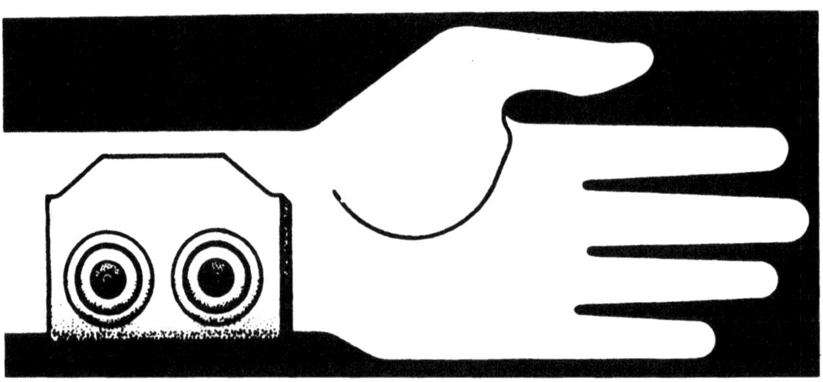

Fig. 59. A dual pre-gelled self-adhering electrode type for the stimulation of peripheral nerves.

10.2 Procedure for locating the exact stimulus site

1. A bar stimulator cable that comes with every device intended for the measurements of SEPs is plugged into the stimulator.
2. With the patient's hand palm-up, a neutral relaxed position is obtained. Relaxation is important for obtaining good SEPs.
3. The inner surface of the wrist is wiped with alcohol to remove skin oils.
4. The area then has to be dried completely, as otherwise alcohol will break down conductive gel.
5. A small amount of conductive gel is placed on each disk of the bar electrode.
6. Location of the nerve to be stimulated (see also Figs. 76, 77):
   - *Median Nerve.* Lateral to the tendon that runs down to the wrist towards the thumb.
   - *Posterior tibial nerve.* The inside of the ankle between the medial malleolus and the Achilles tendon.
7. The stimulating bar is placed over the median or posterior tibial nerve, cathode proximal.
8. Explain the patient that he or she is going to feel some pulses at the wrist/ankle which will begin at low level and increase until the thumb/toe twitches.
9. A toggle at the stimulator box is depressed to the 1 sec position.
10. Thereafter the intensity knob for current is turned on and increased slowly until the patient can feel the pulse. For median nerve the patient should feel the pulse travel down towards thumb and index finger. The pulse should not be felt in the little finger as this is the ulnar nerve. For posterior tibial nerve stimulation the pulse should travel down to the sole of the foot to the big toe. If the pulse is felt over the heel, stimulation is over the wrong branch of nerves. Change the location of the bar stimulator. A light at the stimulator box will flash each time a pulse is delivered.
11. If necessary, the stimulating bar has to be moved in order to find the optimum stimulus location. Sometimes pressure on the bar has to be applied as the nerve is located deep below the skin. Avoid gel from one electrode to connect with the other. This would result in a 'bridge' and shorts out the stimulus.
12. Once a pulse is felt in the correct location and thumb or toe twitches, the stimulus intensity on display should be remembered for later stimulation recording. The toggle switch on the stimulator is switched back to the center position.
13. The stimulating area is now cleaned with alcohol and dried completely.
14. Pre-gelled pediatric ECG disc electrodes are applied to the exact location of the bar electrodes and connected with the stimulator cable.
15. The stimulation electrodes, the cathode, or the negative pole (red), and the anode, or the positive pole (black), are placed 2–3 cm apart.

As the current flows from the anode to the cathode, a negative charge is accumulated at the cathode near the nerve surface resulting in depolarization. The positive charge accumulated at the anode near nerve surface leads to hyperpolarization

Fig. 60). Therefore the cathode has to be placed close to the recording site as not to diminish the amplitude of nerve volleys that travel along the nerve pathways.

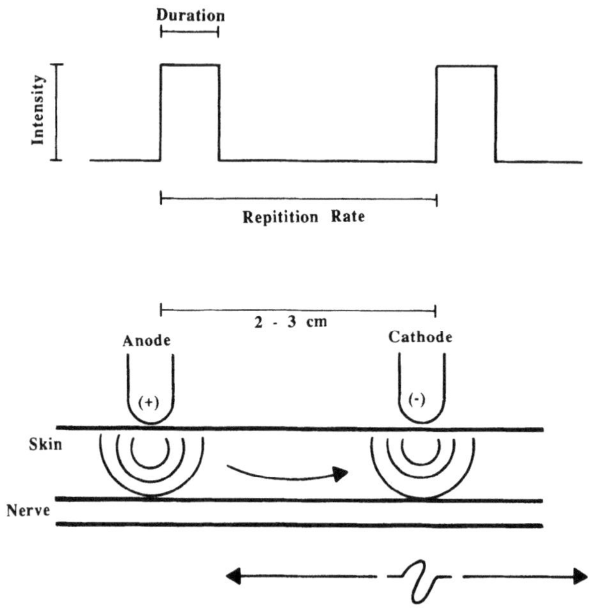

Fig. 60. The two prongs at the stimulation site, where the cathode (red) is located proximally.

10.3 The stimulus necessary for SEP recording

The stimulus is of a square wave nature with a preset duration, amplitude and repetition. For SEP recording:

a) a duration up to 100 ms,
b) an amplitude of 5–10 Volt constant current with 5–30 mA, and
c) a repetition rate between 5–10 Hz are selected.

A visible and maximal motor twitch has to be elicited, which has the following classifications:

a) subthreshold – no axons fire
b) threshold – some axons fire
c) sub-maximal – more axons fire
d) supra-maximal – all axons fire

Voltage and current necessary for each individual may vary greatly. Sub-maximal motor stimulation is usually enough to fire all of the sensory axons. The most important consideration for adequate stimulation is the stimulation over a large nerve and the eliciting of motor responses. The information, therefore, is complex with inputs reaching the central structures at different times [74].

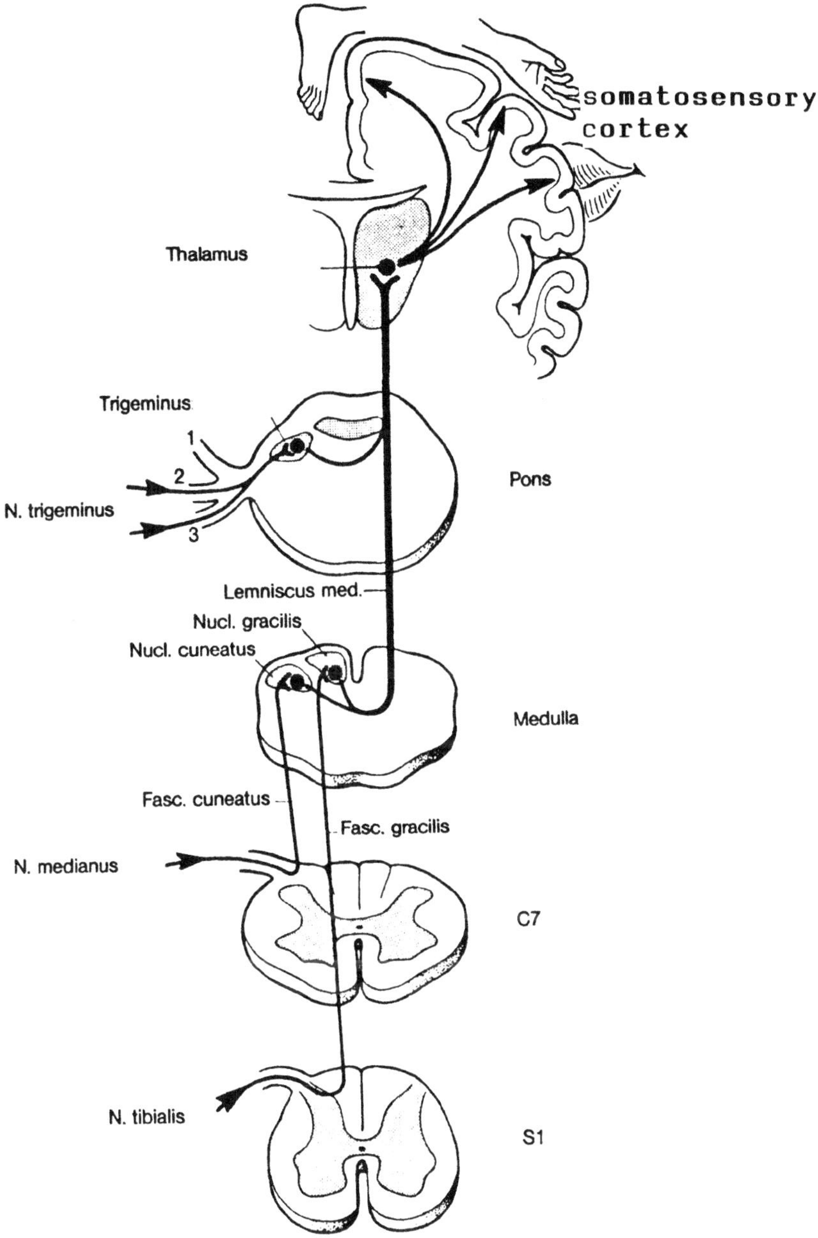

Fig. 61. Pathway of sensory information from the skin to the cerebral cortex (adapted from Stöhr & Petruch [76]).

In order to reduce stimulation artifacts which may jolt the preamplifier input, the following facts have to be taken into consideration:

a) isolate the stimulus from the recording electrodes with a separate transformer,
b) keep the stimulator away from the recording electrodes,
c) remove perspiration from the stimulating area,
d) keep recording and stimulating electrode impedance low,
e) keep the recording leads away from the stimulator as they work as antennas,
f) place a ground *between* stimulator and the recording electrodes.

From the point of origin, the stimulation site, volleys of impulses travel along the peripheral nerves, the nerve plexus, and enter the dorsal column of the spinal cord. From there the ascending lemniscal pathways take the stimuli to the thalamus and the primary sensory cortex (Fig. 61) [75]. The ascending volley can be recorded from any of the several sites along this pathway [77].

10.4 Recording electrodes

The recording of the SEP consists of a pair of electrodes. One of the electrodes is placed over the anatomical site which corresponds to the location along the sensory nerve pathway which is most likely to detect an impulse. This is called the active site. The other electrode is placed in a location not involved in the pathway response. This is called the referential site.

As electrodes are placed on specific scalp locations to obtain a maximum response to a sensory stimulus, the *international 10/20 system* measurement, the

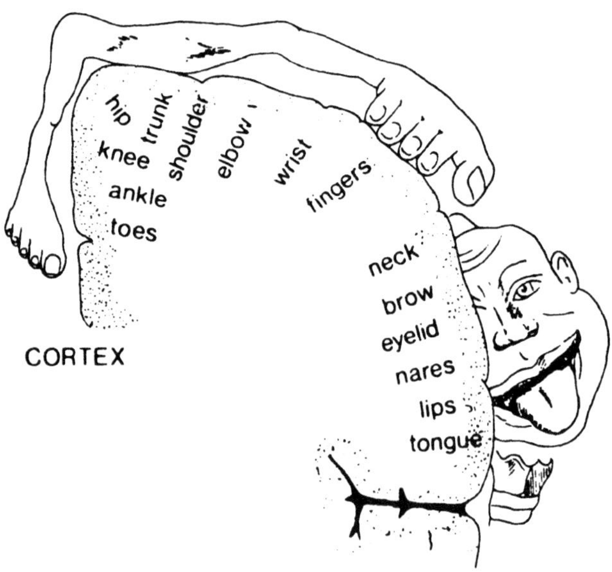

Fig. 62. Localisation of the site of input to the primary sensory cortex.

standard of neurological practice, is used. Measurements are made using bony landmarks on the skull as reference points. With this technique it is possible to consider interindividual differences in circumference and shape of the skull and find the location on the skull which is above the corresponding area of the stimulated part of the body (Fig. 62).

10.5 Procedure for head measurements to determine electrode location

1. Locate the nasion and the inion. The nasion is the bony ridge between the eyes. The inion is the bony ridge at the back of the head.
2. Measure the distance in cm between the nasion and the inion over the top of the head.
3. Mark with a grease pencil 50% of the nasion-inion distance (= C_Z, the vertex) and 2 cm behind the vertex (C_Z') (Fig. 63). Markings are perpendicular to the tape.

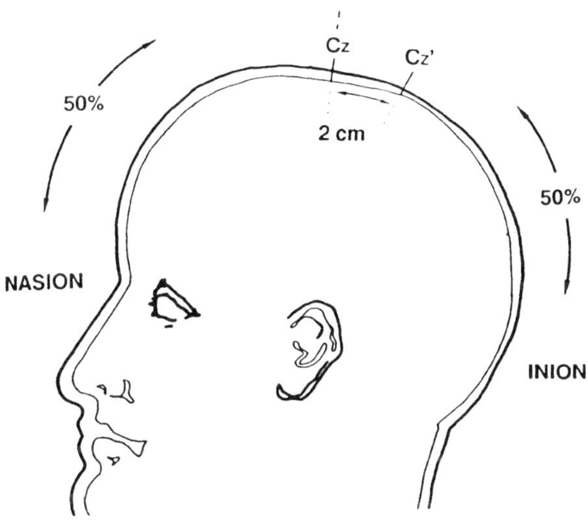

Fig. 63. Localisation of the Vertex (C_z) and the C_z' point.

4. Locate the pre-auricular points. The pre-auricular points are on each side of the head immediately in front of the ear channel.
5. Measure the distance between the pre-auricular points going through the vertex mark.
6. Mark 50% of the pre-auricular distance perpendicular to the tape crossing the vertex mark C_Z. Leaving the tape in place on either side of the vertex mark (C_Z), measure down 20% of the total preauricular distance. Make a mark both perpendicular and parallel to the tape. These marks are called C_3 on the left and C_4 on the right-hand side.

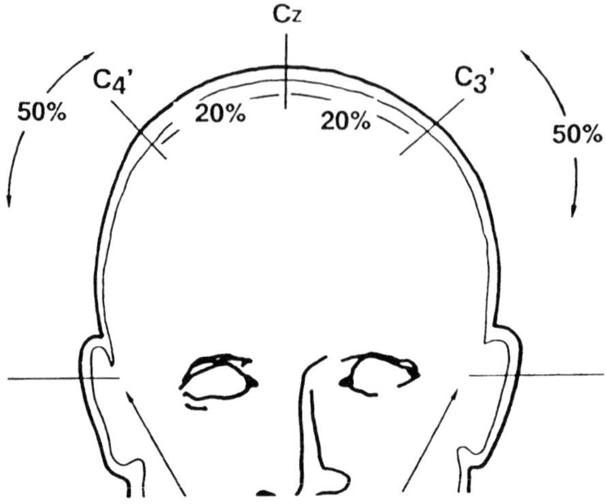

Fig. 64. Determination of C_3' and C_4' for positioning of the active electrode on the scalp.

7. Measure 2 cm behind C_3 and C_4 at the same level. These marks are called C_3' and C_4' on the left and right-hand side respectively (Fig. 64).
8. The other electrode site on the scalp is the middle of the forehead. This point is called FpZ. Measuring is not necessary since it is the reference and exact location is not critical.
9. There are now 4 electrode locations (FpZ, C_z', C_3', C_4') on the scalp for recording the SEP cortical potentials. They are used in various combinations depending on the nerve being stimulated (Fig. 65). The list below will show which electrode location will be used at the specific nerve stimulation.

Recording sites at different nerve stimulation
FpZ-reference for right or left median nerve stimulation
C_3' – for right median nerve stimulation
C_4' – for left median nerve stimulation
C_Z' – for right or left leg stimulation
FpZ – reference during left or right left stimulation

10. Additional electrodes may be placed on Erbs point (EP) bilaterally for median nerve stimulation (midpoint of the supraclavicular area = *plexus brachialis*) and over the popliteal fossa (PF) bilaterally for posterior tibial nerve stimulation. Erbs point is located just above the clavicle and medial to the point of attachment of the sternocleidomastoid muscle (see also Fig. 67). The popliteal fossa is located on the back of the leg at the bend of the knee. The electrodes are placed at the popliteal fossa (PF) with the reference 2 cm below.
11. Still further electrode placements may be used to evaluate the functional integrity of a signal in the spinal cord, i.e. the C_7 and C_2 locations (above the proc. spinosi of the 2nd and 7th cervical vertebra) or the proc. spinosi of the lumbar vertebra (see also Fig. 70).

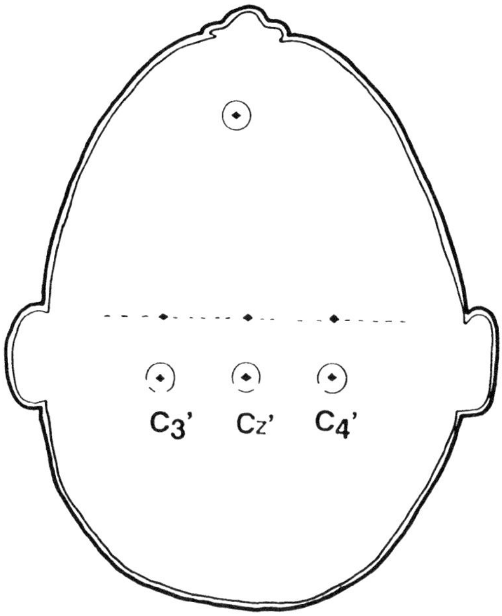

Fig. 65. Electrode locations on the scalp for recording of cortical response during SEP stimulation.

10.6 Accessory for electrode placement and removal

The following is a list of materials necessary for the application and removal of the cup electrodes using collodion glue.

1. *For measuring the head:*
   a) tape measure with centimeter markings, and
   b) grease pencil.
2. *For the application of recording electrodes*
   a) cup electrodes (silver/silver chloride, gold). These electrodes have a doomed top with a small hole in the centre suitable for the application of electrode gel. A highly flexible PVC covered cable is attached to the electrode by a crimp connection. The joint is covered by shrink sleeving,
   b) collodion,
   c) alcohol/acetone prep pads,
   d) a 5 cc syringe with a blunt tip stub adaptor filled with conductive gel, and
   e) an air compressor (optional) including hose and applicator tip. This item is designed as an electrode positioning instrument, an air gun, and as a collodion tube piercer and storage unit. The applicator is provided with a spigot which fits into the hole of the electrode (Fig. 66).
3. *For the application of stimulation electrodes:*
   a) bar stimulator to locate site,
   b) ECG electrodes,

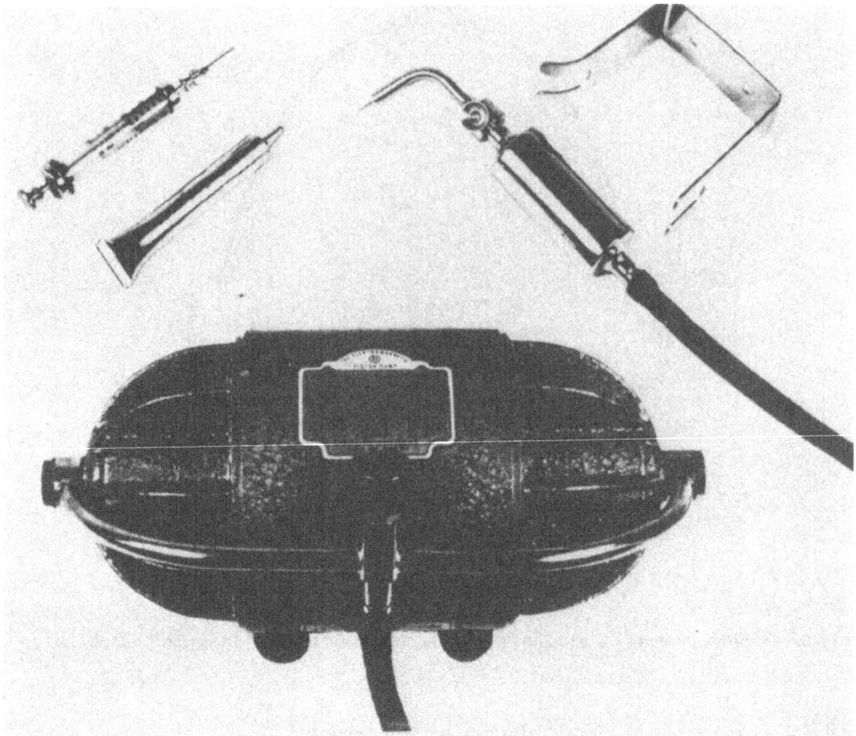

Fig. 66. Electrode applicator and reciprotor pump. The latter is a silent compressor suitable for the provision of drying air to rapidly set collodion.

   c) ground plate and strap, and
   d) conductive gel.
4. *Items for removing the electrodes:*
   a) acetone, necessary to remove collodion,
   b) cotton balls, and
   c) gauze squares.

When mounting in surgical situations, it is advisable to monitor two anatomical sites along the nerve pathway at risk. In one channel the *active* electrode site is over a peripheral nerve such as Erbs Point in median nerve stimulation or the fossa poplitea in posterior tibial nerve stimulation. In the other channel, the active electrode is placed over the sensory cortex site which corresponds to the nerve being stimulated (Fig. 67). Generally, the peripheral nerve response serves as a control. If, for example, only the cortical response is lost it is likely due to the surgical procedure. If both responses are lost it is most likely due to improper stimulation.

The *inactive* electrode is placed on a point that is minimally involved in the sensory pathway. No electrode site on the body, however, will be entirely inactive. Locations which may be used for the inactive electrode include the forehead (FpZ), the mastoid, the chin, the shoulder, or a leg.

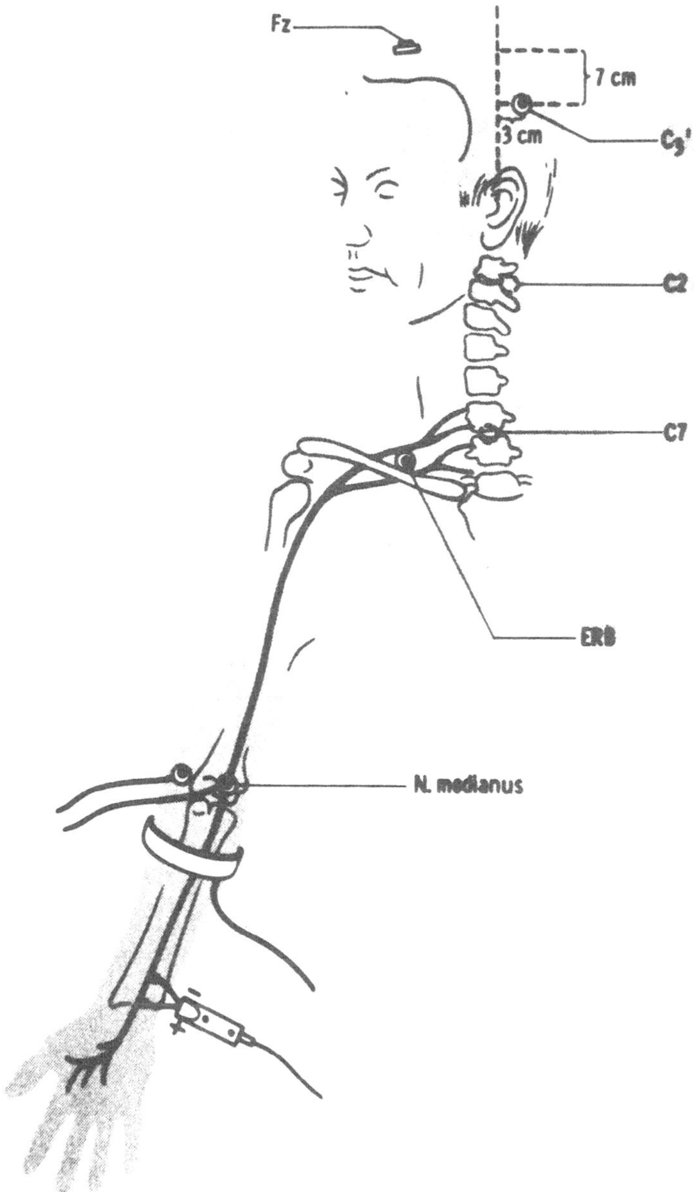

Fig. 67. The various points along the sensory pathway which can be used for recording of median nerve evoked potentials and the evaluation of functional integrity.

A large *plate electrode* is used as a common or ground reference. This electrode is usually placed around the limb being stimulated, *between* the stimulation site and the first recording electrode. The ground electrode is necessary to *improve* the ability of the amplifier to reject noise. Its main purpose is:

a) to maintain a low impedance level, and
b) to be located near the stimulation site.

As the recording electrodes pick up the evoked response, they in turn are connected to the differential preamplifier. After all noise is filtered out, the sample is converted into digital form and averaged together with other samples. Finally, the computerized information is presented on display. This entire sequence occurs within microseconds (Fig. 68).

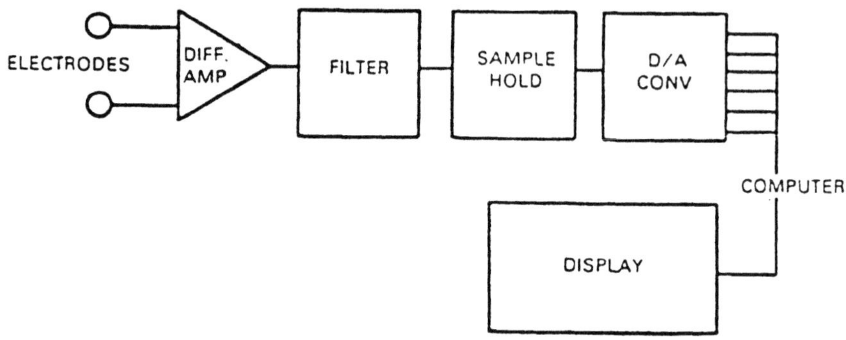

Fig. 68. Schematic drawing of the operation of equipment during evoked potential monitoring.

Similarly, as during power spectral analysis of the EEG, the *electrodes are fundamental to collect a clear evoked response*. The tiny currents of the evoked potential must be picked up in order to flow through the electrodes. Currents will only flow through electrodes to the extent that this path offers *less resistance* than the skin path. Therefore, the aim is to *minimize* as much as possible electrode impedance at the electrode connection to the skin. This can be achieved by observing meticulously each of the following steps:

1. Part the hair at the recording site.
2. Remove the overlying dead skin (stratum corneum) from the recording site by abrading it with 'Omniprep™' in a circular motion.

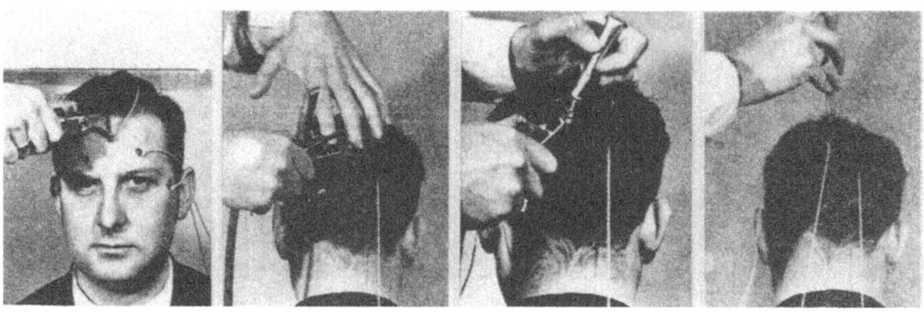

Fig. 69. Application procedure to apply the 'stick on' electrodes to the scalp for evoked potential recording.

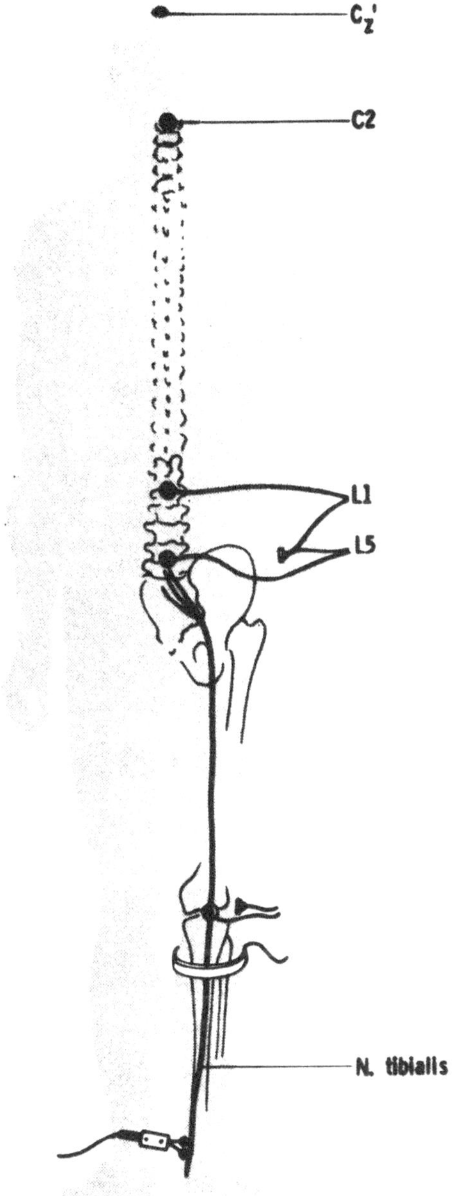

Fig. 70. Position of the ground electrode on the leg for posterior tibial nerve stimulation.

3. Remove perspiration and oil (as this spreads the potential, decreases amplitude and increases artifacts) by gently brushing with alcohol.
4. Let recording site dry completely, then tape the electrode firmly to the skin surface with the electrode wire orientated towards the back of the head.
5. Place the tip of the electrode applicator through the hole in the electrode cap to hold the electrode in place and prevent collodion from getting under the electrode (Fig. 69).
6. Squeeze a moderate amount of collodion glue around the flange of the cup electrodes and spread it over the entire surface and the adjacent skin.
7. Dry collodion thoroughly using compressed air. After the collodion has been applied around the electrode, blow air from the applicator tap around the flange of the electrode, rapidly drying the collodion and fixing the electrode to the scalp.
8. The cup of the electrode must now be filled with conductive gel using a clean, blunt-tipped specially fashioned needle point causing the scalp to be abraded whilst applying the electrode jelly through the hole of the electrode. The skin is being abraded lightly by rubbing back and forth.
9. The ground electrode should next be applied around the arm between the wrist (the stimulation site) and the elbow for median nerve stimulation, or around the leg between the ankle (the stimulation site) and the knee for posterior tibial nerve stimulation (Figs. 67, 70). If both nerves are stimulated, the arm may be used for grounding.
10. Before applying the ground electrode, the area should be thoroughly wiped with alcohol to remove oil. Thereafter, the site should be dried completely with a dry gauze pad.

Conductive gel for the metal ground plate, or physiologic saline solution (Fig. 71),

Fig. 71. The two different types of ground electrodes used in evoked potential monitoring. *Top*: Velcro ground strap electrode. *Bottom*: Ground plate electrode.

are applied to the electrode which is secured around the limb with a velcro strap, making sure that the cable emerges from the electrode towards the head.

Subdermal needle electrodes may also be used for recording. However, different types of electrodes should *never* be mixed in order to avoid DC potentials. The main reason why this type of electrode for EEG and Evoked Potential recording is seldomly used is that these electrodes have to be autoclaved in order to avoid any possibility of an infection. Additionally, normal plastic insulation does not withstand repetitive autoclaving at 260–270 °F so that the needle electrodes are considered the most unreliable and expensive. For short-term recording, however, when the patient is not moving, one set of needle electrodes made of special platinum alloy could be used.

10.7 Connection of electrodes with the preamplifier

At least two channels of input can be monitored with every instrument. Thus, the following scheme will outline in which manner the electrodes are connected to the preamplifier input in order to get an appropriate signal.

	Channel 1		**Ground**	**Channel 2**	
	(+)	(–)	(R)	(–)	(+)
Median Nerve stimulation					
– Left	FpZ	C_4'	Left arm strap	Left Erb	FpZ
– Right	FpZ	C_3'	Right arm strap	Right Erb	FpZ
Posterior Tibial Nerve Stimulation					
– Left	FpZ	C_z'	Left leg strap	Left popliteal fossa	2 cm below
– Right	FpZ	C_Z'	Right leg strap	Right popliteal fossa	2 cm below

Before one gets into the operating mode, it is important to undertake an impedance check in order to avoid any lead problems and correct the electrode site if necessary that may show an impedance of 5 kOhm or more. For proper recording, an impedance around 2 kOhm should be attempted and be equal for all electrodes.

10.8 Trouble shooting to eliminate high impedance and electrical noise

1. If the impedance is too high in all electrodes, check the ground as it affects all other electrodes.
2. If the impedance is greater than 5 kOhm for any of the electrodes, reinsert the stub adaptor into the hole of the electrode dome, abrade the skin by turning motions and insert some more conductive gel or paste, so as to completely filling out the space between electrode dome and skin.
3. Replace the electrode if the impedance still stays high after attempting to abrade the skin for at least twice.

It is necessary to reduce the impedance as much as possible for the following reasons:

1. A high impedance increases noise pickup, since this electrode with its cable will act as an antenna for all sorts of tiny electromagnetic waves in the air.
2. A high impedance will generate random noise by itself.

Current will flow through the recording electrodes only if this path offers *less resistance* than the surrounding skin. Therefore, it is mandatory to minimize impedance at the electrode connection to the skin as much as possible in order to get a signal from the underlying cortex. Transmission of activity through the membranes covering the cortex, the cerebrospinal fluid, the bony skull, and the skin all result in a diffusion of potential changes which ultimately will result in a decrease of amplitude of the evoked potential and a blurred spatial distribution. Thus, in evoked potential testing, it is always necessary to make the signal as large as possible and to reduce the noise as much as possible. The objective is to minimize noise *before* it gets into the instrument. Otherwise noise must be reduced within the instrument via filters, signal averaging, etc. The following types of noise should be avoided as they disturb the active signal significantly:

1. Exogenous noise from the environment such as motors (i.e. the respirator), lights in the room, unproperly grounded ECG and/or blood pressure monitors, TV screens, 50–60 cycle stimulus artifacts, and sounds from the loudspeaker.
2. Endogenous noise within the patient such as ECG, ERG, EMG, blinks, and muscle twitch.

The solution to reaching lower electrode impedance, is in the first case, to remove the exogenous source of noise and, in the second case, to relax the patient. Patient relaxation is important to obtain good evoked potentials signals. Sometimes a change in the position of the preamplifier will result in lesser interference with exogenous sources of noise while in another case the machine which generates the noise, such as the respirator, should be positioned further away from the patient. At times it is also helpful to make sure that the external sources are properly grounded via a third electrode wire. In order to avoid active-to-reference jumper, it is advisable to connect the reference electrode with the ground as the ground is *not a signal path* and not an earth ground. Faraday screens are not really a necessity.

Another potential source of electrode noise are the wire leads of the active, reference, and ground electrodes. One should keep the following advice in mind:

1. All electrodes are to be made of the same material in order to avoid DC potentials.
2. Leads may act as conductors, working like antennas which pick up electrical noise from the environment. Thus, they have to be kept away from sources of noise and should not be crossed by other leads (e.g. ECG). All leads follow the *same path* to the preamplifier, should be of the same length, and should be *kept as short* as possible.
3. A *broken protection* insulation in any part of the cable will result in artifacts. Once the signal is picked up from the electrode and transmitted to the preamplifier via cables, the latter has a number of options to deal with the incoming signal.

10.9 The differential amplifier for EEG and the EP measurements

The differential amplifier capabilities:

1. It amplifies the voltage difference between active and reference lead (Fig. 72).
2. It rejects similarities between active and reference input (=common mode rejection).
3. A positive signal at the active input will be displayed downwards.
4. A positive signal at the reference input will be displayed upwards (Fig. 72).
5. The differential amplifier is important for noise rejection insofar as
   a) the impedance of electrodes is lowered,
   b) the impedance levels between the electrodes is balanced
   c) it aids in evaluating the choice of montage, and
   d) it selects the appropriate amplifier gain for the signal.

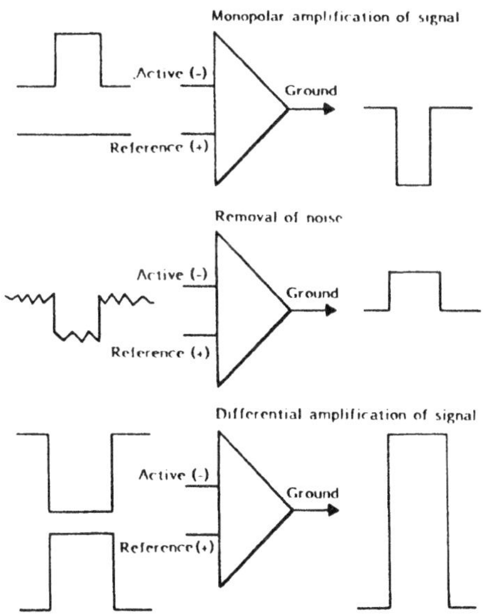

Fig. 72. Principle of the amplifier, what it is and how it works in recording evoked potentials.

11. Optimising signal-to-noise ratio

11.1 Introduction

11.1.1 Filtering of noise within the recording machine

Electrical noise and interfering signals contain a random mixture of all frequencies, whereas the evoked potential only contains selected frequencies. The objective in filtering is to remove noise which does not overlap the signal by blocking it from entering the computer system (Fig. 73).

The sharp edges as shown in Fig. 73 (f and f_1) represent the points where noise frequencies are attenuated. However, in reality the filter has a gradual slope. For

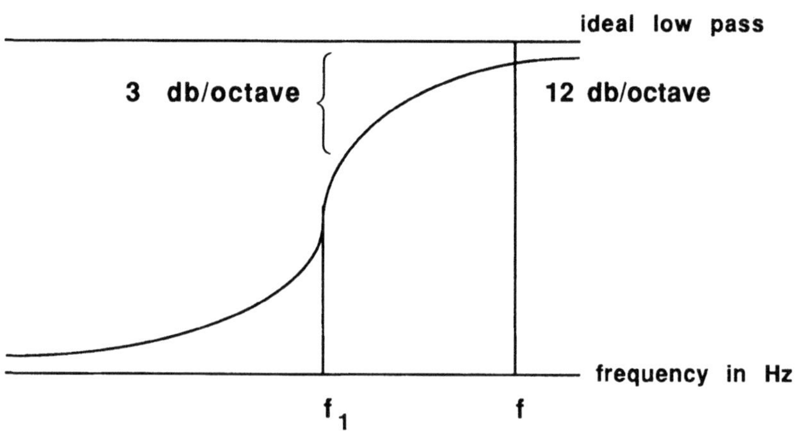

Fig. 73. Basic functioning of a filter to reduce electrical noise in an incoming evoked potential signal.

instance at point 'f_1' = 100 Hz cut-off, noise will still be 1/4 times as large as at 50 Hz. Noise is not totally eliminated by filtering, as an increase in the slope of attenuation otherwise would result in wiggle of the base line. Thus, a filter necessary for the analysis of a pure signal is limited by its high- and lowcut. If, however, filters would result in a partial removal of the active signal in evoked potential monitoring, a change in the latency and the morphology would result.

Analog to digital conversion. The process of digitization is best described by the sequence: sample, hold, digitize, and label. In discrete steps of time the sampling is fast enough not to miss any change in the spectral event. Thus the *higher the sampling rate*, the better the *resolution* and the smoother the changes of the wave. For instance a unit like the Neurotrac™ has a sampling rate of 2.56 kHz/channel; the Lifescan™ Brain Activity monitor, on the other hand, has a sampling rate of 3.1 kHz/channel which results in better resolution and smoother wave display.

11.1.2 The process of averaging

The evoked potential picked up from the scalp ranges from 1–6 μV compared to EEG potentials in which the amplitude ranges from 10–200 μV. For this reason special techniques need to be employed to extract the evoked potential wave from the ongoing background EEG. This technique is called signal avaraging. The EEG is a continuous signal, while the evoked potential is a specific, short duration response to the stimulus. Thus the evoked potential is contained in the raw EEG signal. In averaging, the EEG activity during a specified time period immediately following the delivery of the stimulus is analyzed. For analysis, the continuous EEG signal is converted into a series of discrete numbers, each number being proportional to the amplitude of the EEG at that point. The numbers are then stored in specific memory bins in the computer. This is called analog-to-digital conversion. With the delivery of the next stimulus the EEG signal during the specified time period is again digitized, and each point summed with the value previously stored in the corresponding bin, and then stored back into that bin.

The general principle of averaging is that if a sum of values is taken that are essentially random (in this case the EEG signal is considered to be random) the digitized values will be both positive and negative values, and their sum will approach zero. The nervous system, however, responds to a stimulus essentially in the same manner each time, and the sum (average) of these digitized values in each bin will increase (either positively or negatively) to produce a consistent, recognizable waveform. The more samples are used in obtaining the average, the greater the reduction of the background or random noise compared to the evoked potential. Averaging, thus, is a signal enhancement technique in which the amplitude of time-locked neural activity (the evoked potential) is enhanced relative to random activity of the underlying EEG and EMG. As the number of repetitions increases, the effective EEG amplitude decreases, leaving the 'pure' cortical response to the stimulus (Fig. 74).

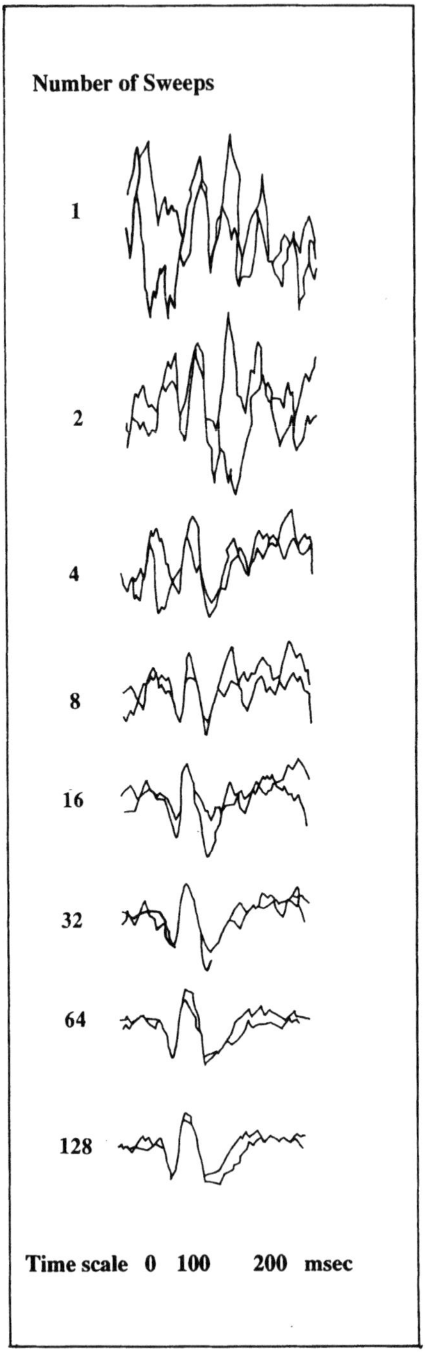

Fig. 74. Principle of averaging in evoked potential monitoring. Note, the higher the number of sweeps being averaged, the better the evoked peak (low signal/noise ratio).

In practice the number of averaging is sufficient:
a) when the signal stops changing on the screen and stabilizes, and
b) the results can be replicated *and are* consistent.

11.1.3 The stimulus rate

The stimulus rate is the number of times the stimulus is applied to the patient per second. The stimulus may be changed according to the preference of the user. The stimulus rate must not be faster than the analysis time. If the patient is stimulated faster than the sampling time of an EEG epoch, a stimulus artifact will be present in the collected data. Most of the units prevent this from happening by limiting the number of stimulus rate options according to the analysis time selected by the user. The longer the analysis time, the *slower* the rate must be, causing fewer options to appear on the menu. The maximum rate of stimulation may be selected when the analysis time is changed to 50 ms. It is important to note that the stimulation rate as it is selected from the menu is offset by some fraction of rate in order to avoid synchronization with 50 or 60 Hz harmonics respectively. The actual values are as follows:

Stimulation rate	Actual (60 Hz)	Actual (50 Hz)	Stimulation rate	Actual (60 Hz)	Actual (50 Hz)
1	0.992	0.990	6	5.714	5.882
2	1.967	1.961	7	7.059	6.818
3	2.927	3.030	8	8.000	8.000
4	3.871	4.000	9	9.000	9.091
5	4.800	4.762	10	10.345	10.417

11.1.4 Stimulus repetitions (sweeps)

The stimulus repetition is a parameter which reflects the total number of stimuli the patient receives to produce an average waveform. This parameter is not dependent on other parameters. Generally, the higher the number of averages (stimulus repetitions), the more noise-free is the resulting averaged waveform. Median nerve and posterior tibial nerve SEPs can usually be obtained using 256 stimulus repetitions or 512 in a noisy environment. Spinal or cervical responses require considerably more stimuli. The values of stimulus repetitions are set on the keyboard of the evoked potential analyzer and may range from 32 to 64–128–256–512–1024–2048 and 4096 sweeps.

11.1.5 The analysis time

The analysis time is the amount in the EEG (in msec) following each stimulation which is collected by the unit and used to calculate the average waveform. The analysis time determines the maximum horizontal time scale able to be analyzed on the display. Thus, the time must be set so that the peaks of interest of the wave fall within this time. If a time is selected which is incompatible with the chosen stimulus rate, a message will warn the user and the rate is changed to be compatible with the analysis time (Fig. 75). The analysis time which usually can be selected ranges from 80 to 100–150–200–250–300–400 and 500 ms.

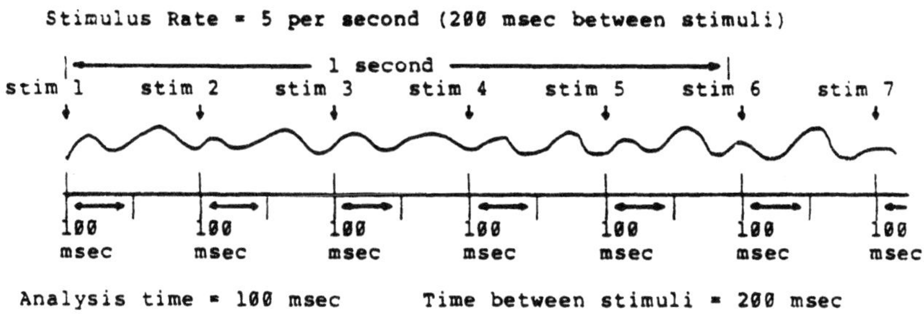

Fig. 75. Stimulus and recording parameters in relation to the raw signal of the EEG. If stimulation is 5/s there is a 200 ms time span necessary between each stimulus (1/5 = 200 ms). The time before another stimulus can be averaged is 200 ms.

The parameters usually chosen for SEP analysis are as follows:

	Median nerve	*Posterior tibial nerve*
Stimulus rate	5/sec	5/sec
Analysis time	50–150 ms	100–300 ms
Repetitions (sweeps)	256 (*or* 512)	256 (*or* 512)

11.2 Procedure to locate the optimal stimulus site

In order to properly locate the optimal stimulus site on the skin overlying the superficial nerve such as in median nerve recording, the patient's hand should be placed palm-up and in a natural relaxed position. The inner surface of the wrist has to be wiped with alcohol to remove any skin oils. Thereafter, the area has to be dried completely with a clean gauze square. The alcohol has to be wiped away as otherwise it will break down the conductive gel of the stimulating electrodes. Now the nerve to be stimulated is located with the additionally supplied bar stimulator. For median nerve location the tendon which runs down the wrist approximately in the middle of the arm is located. The median nerve is usually found at the wrist adjacent to the large tendon which runs from the middle of the arm from the elbow towards the little finger of the hand (Fig. 76).

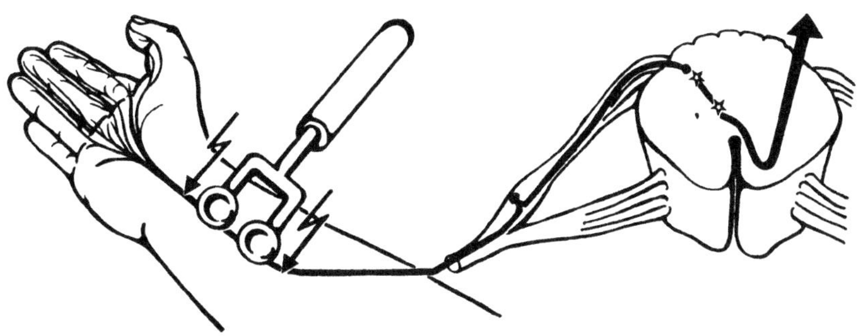

Fig. 76. Location of the median nerve using the bar stimulator.

For posterior tibial nerve location the *inside* ankle, between medial malleolus (ankle bone) and Achilles tendon has to be stimulated (Fig. 77). The patient should always be informed that a selective site is going to be stimulated and he or she will begin to feel some pulses at the wrist/ankle. Begin at very low level and increase gradually until the thumb/toe twitches.

The stimulating bar is now placed over the median/posterior tibial nerve with the cathode (the negative electrode usually colored red) located *proximally*. Then the bar stimulator cable is plugged into the stimulating device and intensity is slowly turned on. The pulses will be delivered at a rate of 1 per second. Usually a pulse light next to the digital display will flash each time a pulse is delivered. While the current is being increased, the patient is asked where he/she feels the pulse. For the median nerve, the patient should feel the pulse travel down into the hand towards the thumb and the index finger. The pulse should not be felt in the little finger as this is the ulnar nerve. For posterior tibial nerve the pulse should travel down the sole of the foot to the toes. If the pulses are felt in the heel, the stimulus is over the wrong branch of the nerve. Then the stimulating bar electrode is moved in order to find the optimal stimulus location. One may exert some pressure on the bar as the nerve sometimes is located deep below the skin. Conductive gel is applied on the site of the stimulating electrodes to prevent burning of the skin. One electrode should not connect with the other electrode as this causes a 'salt bridge' shortening out the stimulus. Once the pulse is felt in the correct location the stimulus intensity is slowly increased until the thumb/toe twitches. The stimulus intensity, as flashed on the display at which the thumb/toe twitches, is noted and remembered for later stimulation. Thereafter the stimulator is switched off, the bar removed and the indented circles which are shown on the skin from the pressure of the bar electrodes wiped with alcohol and dried completely. Thereafter pre-gelled pediatric ECG electrodes are applied on the stimulation site, one on each circle left by the bar electrodes and connected with the stimulator cable. One always has to remember that the *cathode (usually marked red) is located proximally*. The stimulus intensity may now be increased by a few mA when using the ECG pad electrodes, since no pressure is being applied.

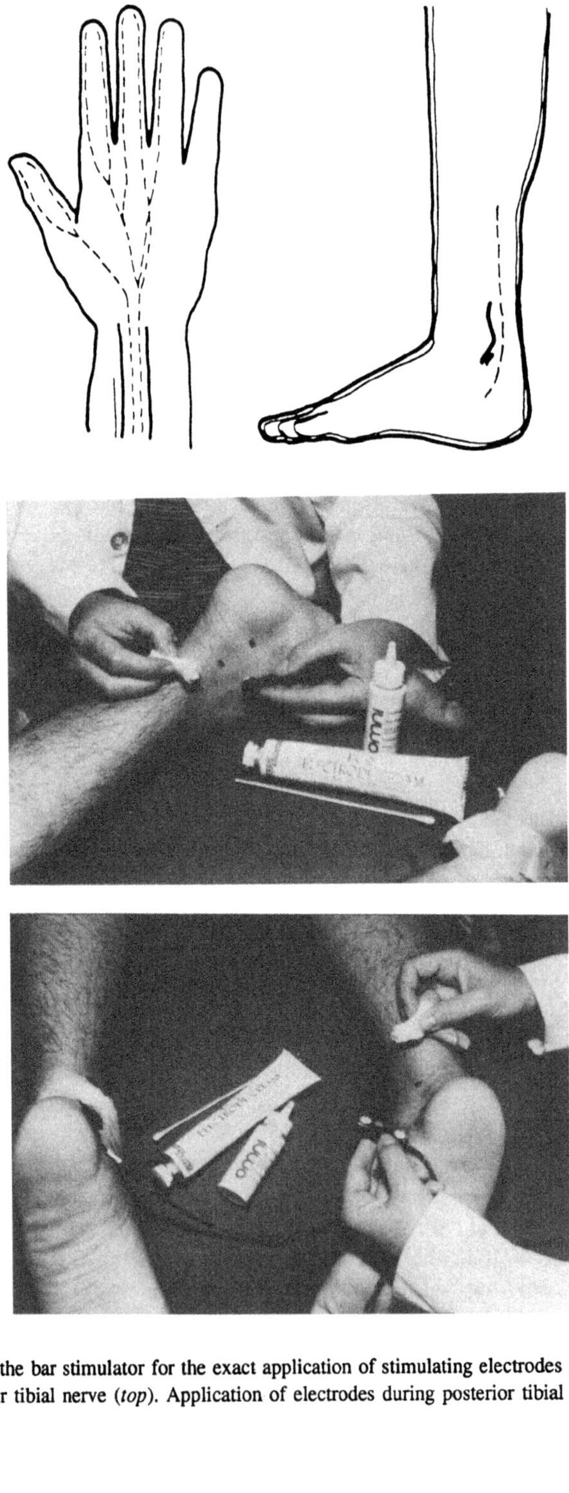

Fig. 77. Use of the bar stimulator for the exact application of stimulating electrodes above the median and the posterior tibial nerve (*top*). Application of electrodes during posterior tibial nerve stimulation (*bottom*).

11.3 Alternating recording and stimulus sites

Somatosensory evoked potentials can be recorded with a varying degree of success from just above any site over the nerve being stimulated. The recording site depends very much on the operation that is being done, and the nerve structure being at risk. Alternate sites of recording include:

a) the cervical cord, usually the skin overlying the proc. spinosus of C_7 and/or C_2 (Fig. 78), and
b) any position along the spinal cord during posterior tibial nerve stimulation. Evoked potentials recorded along the spinal cord are called Spinal Evoked Potentials.

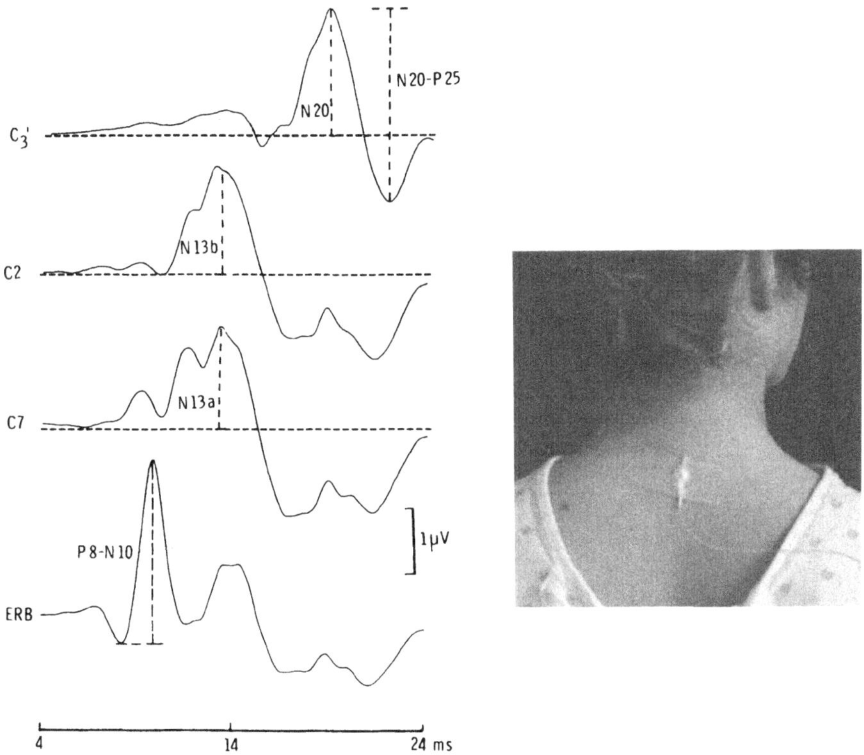

Fig. 78. Alternative sites for recording during SEP monitoring (Erb's point, C_7 and C_2). During median nerve stimulation, position C_7 is useful in recording the entrance of the signal into the spinal cord.

Alternating stimulating sites. The common peroneal nerve as it crosses the head of the fibiula is another common site for stimulation. Sometimes it is impossible to stimulate at the location specified above. Alternative sites may be found by moving up the nerve to the next *superficial* location. For example, if the median nerve is inaccessible, the stimulating electrodes may be placed in the antecubital space or the popliteal fossa in posterior tibialis recording.

Alternating recording electrodes. Besides silver/silver chloride cup electrodes, needle electrodes can be used successfully in the operating room. The reason for their use is the ease of their application. However, when using needle electrodes, special care has to be taken to secure the electrodes with tape to the skin and avoid any form of strain so as not to pull the electrodes out. Needle electrodes can also be used to record from such places as Erbs point which can also be used as stimulating electrode site. The stimulus current applied during the use of needle electrodes, necessary to obtain muscle twitch is *much less* than the current usually necessary with pre-gelled pad electrodes, as the skin acts as an insulator.

Patient clean up. The following procedure is recommended for removing electrodes (stimulating or recording) from the patient:

1. Soak a cotton ball with acetone.
2. Hold the cotton ball into a gauze square and wait about 15 seconds for the acetone to absorb into the gauze.
3. Wipe the gauze across the electrode, allowing removal of the electrode and the gauze from the scalp.
4. Wipe the remaining collodion, electroconductive gel, and acetone with a clean, dry gauze pad until the site is reasonably clean and dry.
5. Repeat the above steps for each electrode attached with collodion.
6. Remove the stimulating disk electrodes.
7. Remove the ground strap(s).
8. Wipe off remaining electro-conductive gel at electrode sites with alcohol.
9. Recheck the scalp for remaining collodion; remove using a gauze with acetone.

12. Evaluating the response of the evoked potential

As somatosensory evoked potentials are transmitted by peripheral nerves, nerve plexus, nerve roots, and dorsal columns of the spinal cord, they follow the lemniscal pathway to the thalamus and the primary sensory cortex. The ascending volley of electrical activity can be picked up from any of the several sites along this pathway. A characteristic waveform is generated as the impulse passes under each of the active electrodes placed along the nerve pathway. The resulting waveforms are named according to the polarity of the signal as it is generated by the body (N for negative, and P for positive), and the approximate time (latency in milliseconds; ms) it takes for that response to occur after the stimulus.

12.1 Median nerve evoked potential

Conduction velocities in peripheral nerves can be recorded by measuring the nerve action potentials that arise in nerve trunks and in the brachial and lumbar plexus. These early EPs are important as they conform an adequate sensory input when the function of higher CNS centers has to be evaluated. A negative potential recorded from the skin overlying the second cervical spine will appear about 14 ms post median nerve stimulation and is thought to arise in the dorsal column nuclei. A typical evoked potential as recorded over the specific sensory cortex, contralateral to the site of stimulation, is seen in Fig. 79. It shows traces obtained from stimulation of the median nerve at the wrist.

The first electrode detecting the impulse is over the brachial plexus at Erbs Point, generating a nerve action potential, generally called N_9. The impulse continues, entering the spinal cord and ascends in the dorsal columns to synapse in the brainstem nuclei.

A second neuron originates in the brain stem and travels to the thalamus, where it synapses.

A third neuron originating in the thalamus projects to the sensory cortex resulting in the waveform detected by the electrode placed over the sensory cortex. The thalamus/thalamocortical radiation response is initially negative, called N_{19}, and is followed by a large physiologically positive potential, called P_{22}.

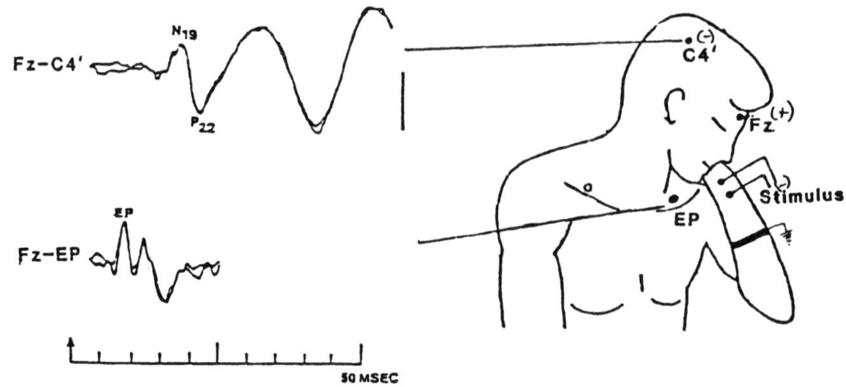

Fig. 79. Normal median nerve SEPs recorded at Erb's point (EP) and at the scalp.

Some observers postulate the N_{19} wave to originate in either the thalamus or the thalamocortical radiation [77]. By convention this wave is labelled N_{20} where the letter designates the polarity and the number indicates the minimum poststimulus latency of 20ms. All observers, however, agree that the preceding waves seen after stimulation and which are distributed widely over the scalp are *subcortical* in origin (Fig. 80–82) [79, 80].

Table 4 summarizes the purported generators of short-latency somatosensory evoked potentials which play an important part in evaluating the function of deeper brain structures.

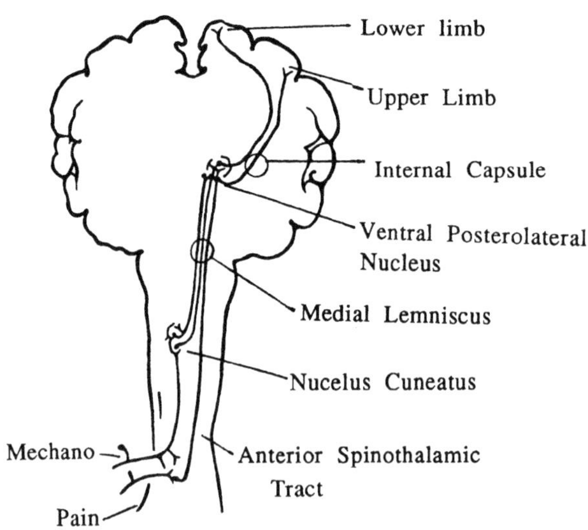

Fig. 80. The central pathways which can be evaluated in somatosensory evoked potential monitoring.

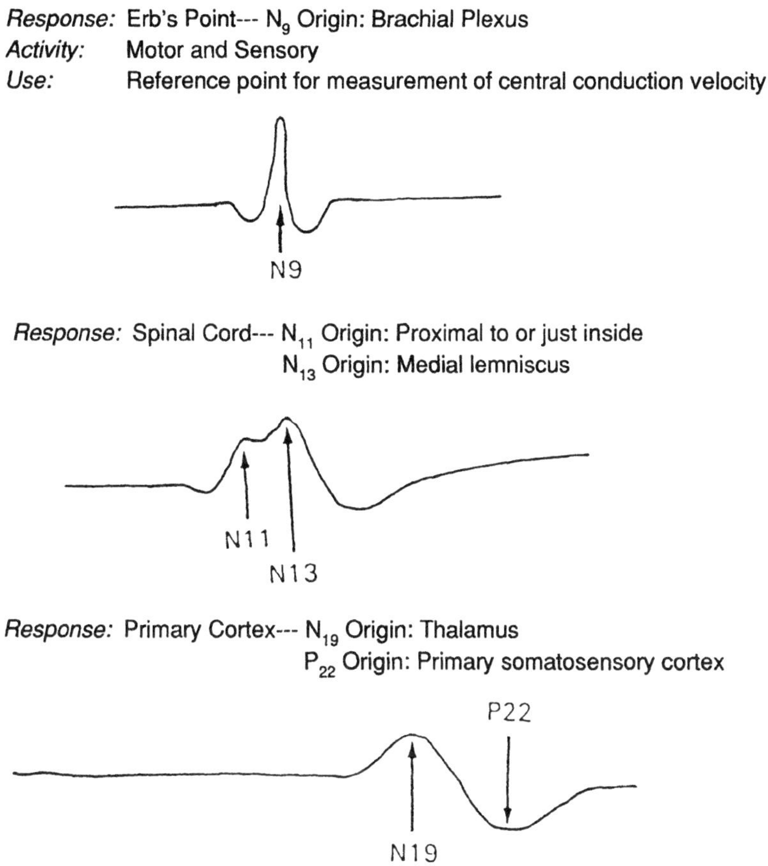

Fig. 81. The various peaks as they are generated during median nerve evoked potential monitoring.

Table 4. Purported generators of short-latency SEPs.

Peak	Generator
N_9	Brachial plexus
N_{11}	Spinal roots of dorsal horn
$N_{13} + N_{14}$	Spinal grey matter of dorsal columns
N_{20}	Primary sensory cortex

* Adapted from Grundy [81].

The non-specific response of the later evoked potentials (over 50 ms) are generated in the primary sensory cortical area and their associative centers.

Pathologic changes at any point along the conducting pathway from the site of stimulation to the specific somatosensory cortex can affect the SEP [82]. A localization of lesion is often possible if early peaks are present and stable, but later peaks are absent or abnormal. Thus SEPs serve to detect pathology, localize the deficit, and in a limited way gauge severity and character of the disorder (Fig. 82).

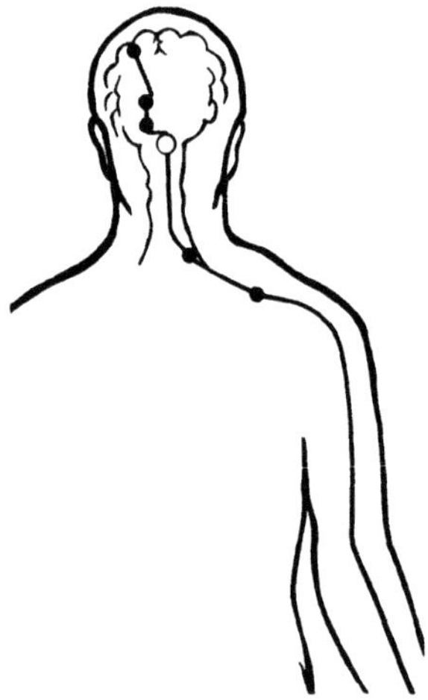

• Generator location

Location of potential disorder	PN	EP	N_{11}	N_{13}	N_{19}	P_{22}	IPL
Peripheral nerve	Ab	Ab	Ab	Ab	Ab	Ab	Stim-EP
Brachial plexus	N	Ab	Ab	Ab	Ab	Ab	Stim-EP
Spinal root	N	N	Ab	Ab	Ab	Ab	EP-N_{11}
Brainstem	N	N	N	Ab	Ab	Ab	N_{11}–N_{13}
Thalamus	N	N	N	N	Ab	Ab	N_{13}–N_{19}
Cortex	N	N	N	N	N	Ab	N_{19}–P_{22}

Abbreviations: PN = Peripheral nerve; EP = Erb's point; Ab = Absent *or* Delayed; N = Normal; Stim = Stimulus; IPL = Inter peak latency increased; N_{11}–P_{22} = Peaks recorded at the corresponding anatomical sites.

Fig. 82. With median evoked potentials it is possible to localize the site of a possible functional disorder in the somatosensory pathway by the absence (Ab) or delay of the peak at the corresponding recording site.

12.2 The posterior tibial nerve evoked potential

Because of the greater distances SEPs, elicited by stimulation of the posterior tibial nerve at the ankle, have greater poststimulus latencies than the SEPs generated by stimulation of the median nerve at the wrist. While the initial negativity at the cortex appears approximately 20 ms after median nerve stimulation, latency of the initial negative peak after posterior tibial nerve stimulation will appear at about 30–40 ms.

In recording the posterior tibial nerve evoked potential some technical problems arise if the forehead reference is used. The length of the sensory nerve pathway from the ankle to the brain introduces a lot of 'artifacts' from the electrical fields generated by other parts of the body. This makes it more difficult for the small potentials to be recorded as they are generated in the sensory nerve pathway. In order to record the peripheral nerve action potentials, the 'active' electrode is placed over the popliteal fossa (PF) at the knee ipsilateral to stimulation and the 'inactive' electrode is placed below and slightly to one side of the PF about 3 cm away.

The Fig. 83 below shows the traces from stimulation of the posterior tibial nerve at the ankle. As the impulse travels up the leg, the characteristic nerve action potential of waveform A is generated as the impulse passes under the electrode at the popliteal fossa, usually designated N_8. The impulse continues, entering the spinal cord, ascending towards the brain in the dorsal columns to the sensory neuron synapses in the brainstem. A second neuron originates in the brainstem and ascends to the thalamus. From the thalamus a third neuron originates and is projected to the sensory cortex. The thalamus/thalamocortical radiation potential (wave B in Fig. 83) is initially negative and called N_{33}. This is followed by a large positive potential, called P_{37}, with another large negative potential following, called N_{45}.

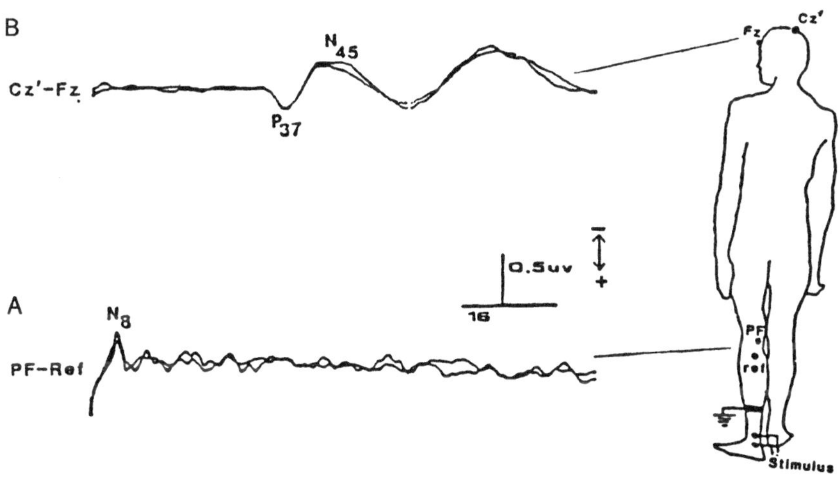

Fig. 83. The normal SEP as it is derived from peripheral and central nervous pathways during posterior tibial nerve stimulation.

The SEPs during posterior tibial nerve stimulation reflect the arrival of nerve volleys at specific landmarks along the sensory pathway resulting in interruptions and delays to be detected and localized. Thus, not only the pathology and the localization of a deficit may be identified, but in a limited way, these SEPs also gauge the severity and the character of the disorder (Fig. 86).

Response during posterior tibial nerve stimulation: Over lumbar spine, N_{20} originates in cauda equina and dorsal root entry. It is used as reference point for measurement of central conduction velocity

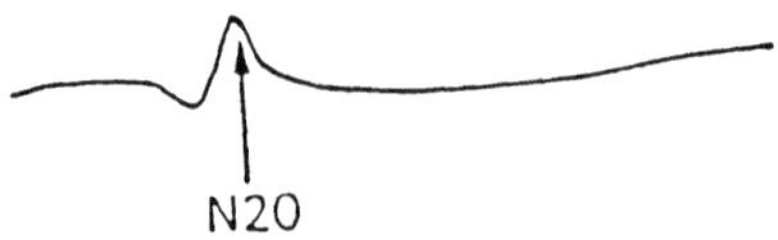

Response of late P_{37}, N_{45}, P_{56}, and N_{74}. Over the cortex P_{37} is generated in the primary somatosensory cortex, and the other peaks originate in higher cortical areas.

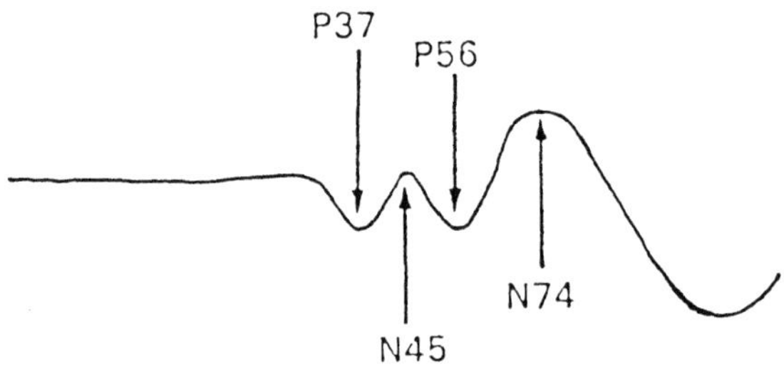

Fig. 84. The various peaks as they are generated during posterior tibial nerve stimulation.

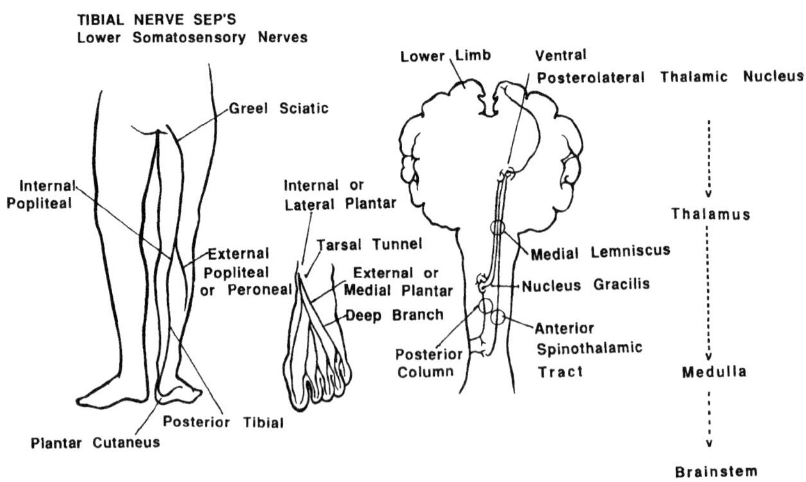

Fig. 85. Physiology of the peripheral and central pathways in somatosensory evoked potential monitoring during stimulation of the posterior tibial nerve.

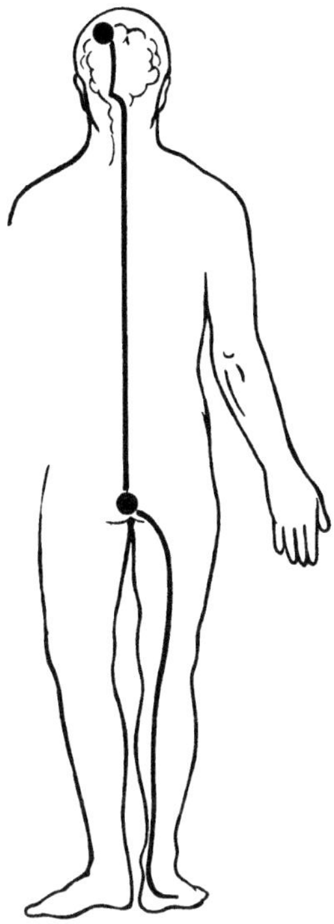

• Generator location

Location of potential disorder	PN	LS	P_{37}
Peripheral nerve	Ab	Ab	Ab
Spinal root	N	Ab	Ab
Spinal cord and brainstem	N	N	Ab

Abbreviations: Ab = Absent *or* Delayed; N = Normal; PN = Peripheral nerve; LS = Lumbar spine; P_{37} = Peak recorded at C_Z' with a latency of 37 msec.

Fig. 86. Localization of the site of functional impairment by recording the peaks at corresponding anatomical sites of the sensory pathway during posterior tibial nerve stimulation.

12.3 Criteria for abnormalities of both median and posterior tibial nerve evoked potential

1. A shift to the right (increase) in latency of the cortical potential, related to the surgical procedure, which does not return to the presurgical baseline when conditions are returned to pre-shift situation.
2. A total loss of the cortical potential when related to the surgical condition.

Once a difference between subject parameter and normal parameter is evident, a functional deficit has to be suspected. The difference may be in regard to an increase in latency or a depression in amplitude of the specific peak reflecting a slowing in transmission speed or a functional deficit in the specific somatosensory pathway. As evoked potentials are a statistical measure, differences which fall within the range of ± 3 standard deviations (= 99%) present *no pathology*. Thus, a 50% change in any of the parameters (latency, amplitude hight) would indicate a pathologic state.

Before interpreting abnormal results one should reconsider and check the following:

1. Is the evoked potential test pathologic in relation to the other diagnostic scheme (blood pressure, anesthetic depth etc.)?
2. Before interpreting the results one has to reconsider the limits of the method; it is only a measure of the functional deficit in the nervous pathway. By no means is it a method intended to give a final diagnosis in regard to the underlying pathology (e.g. tumor, intracerebral bleeding, hypoxemia etc.).
3. The quality of the answer depends very much on the specificity of the question. As non-specific response peaks generally are suppressed first with anesthetic agents, particularly the late negative potential which may *completely disappear*, it is not useful to judge hypoxemia during anesthesia using the late evoked potential. The specific response, however, remains unchanged even in the deeper planes of anesthesia and may even be present in case of barbiturate anesthesia when the EEG has disappeared.

Considerations during evoked potential judgement.

1. Once the waveform is unclear
   a) increase *stimulus intensity* slightly,
   b) decrease repetition rate, and
   c) check electrodes and instrument settings.
2. Once there is too much noise to get a good signal:
   a) increase the number of averages (sweeps),
   b) relax the patient (if awake),
   c) check if other electrical instruments are properly grounded, and
   d) check if electrical cables run over or under the evoked potential electrodes.
3. Check for stimulus factors that might alter the response:
   a) Usually there is a better response on electrical than tactile stimulation.
   b) Any increase in stimulus intensity will result in no change of latency. However, the amplitude increases as intensity increases during thumb twitch.
   c) Filter setting will affect the signal. Thus, the use of a low-cut filter not higher than 10 Hz is recommended. A high-cut filter, however, can go as low as 500 Hz, if necessary. In some devices filter setting is done automatically.
4. Non-pathologic factors that can alter the response:
   a) *Age*: Peripheral nerves age earlier than central nerves; median nerve conduction slows by approximately 0.16 m/sec/year, whereas central conduc-

tion velocity remains stable up to the age of 60 and then slows by 0.78 m/sec/year.

b) *Infants*: The interpeak latency shortens rapidly from age 0–2. Later peak latency shortens slowly over the age 2–6. Adult norms (see pages 137, 138) establish themselves around the age of eight.

c) *Gender*: The latency change for females is shorter by 1 ms than that for males.

d) *Body temperature*: In the limbs, higher temperature will result in a faster conduction. The increase is at a rate of 5% per 1°C. Cooling of the entire body as in cardiovascular surgery, will affect both the peripheral and cortical potentials, with a progressive increase in latency and a decrease in amplitude of all components. With severe hypothermia a total loss of all peripheral and cortical potentials can be expected. With rewarming, a progressive increase in amplitude and a decrease in latency will occur. Cooling of just the spinal cord at the surgical site will not affect the peripheral potentials, but the cortical potential will show a progressive increase in latency and a decrease in amplitude which reverses with rewarming.

e) *Sleep, level of consciousness*: Some effect on the late SEP may be noted.

13. The effect of drugs on the evoked potential

Among the several factors unrelated to surgical manipulation that will cause changes in the evoked response besides cooling of the entire body, anesthetics have the most depressant effect. With the induction of anesthesia, an increase in latency of a few milliseconds is expected in the cortical response. But the peripheral response (i.e. Erbs point, popliteal fossa) should not change. The cortical evoked response stabilizes once anesthesia is maintained.

Some of the anesthetic agents will have a marked depressant effect on the cortex with increasing dosages. Most notable in their effects with increasing concentrations are the inhaled halogen agents (enflurane, isoflurane, halothane, etc), high doses of barbiturates, high doses of benzodiazepines (diazepam), and high doses of narcotics analgesics (fentanyl, sufentanil). The peripheral evoked potential will show no change, but the cortical evoked potential will show a progressive increase in latency, broadening of the complex, and a decrease in amplitude related to the dose of the agent. These effects reverse with a decrease in anesthetic plasma concentration.

Peripherally-acting neuro-muscular blocking agents (curare, pancuronium, atracurium, vecuronium, etc.) will not affect the SEP responses in clinical dosages.

However, local anesthetics, when used for plexus blockade, will prevent the impulse from travelling up the spinal cord as in posterior tibial nerve stimulation, or will block the afferent volleys during median nerve stimulation when axillary blockade is initiated.

When using evoked potentials intraoperatively and in the ICU, one should consider the following:

1. Replicate the response in order to look for consistency and verify the finding. If in doubt, only the replicated response may be used for the evaluation of peak latency and amplitude measurements.
2. For the measurement of conduction velocity:
   - measure the arm length with arm extended to 90° at the shoulder,
   - measure from the negative stimulus electrode to the C_2 vertebra in cm, and
   - calculate nerve conduction velocity in the extremity using the formula:

$$\text{conduction velocity (cm/s)} = \frac{\text{arm length (cm)} \times 10^6}{\text{latency of potential at Erb's point (ms)}}$$

13.1 Application of SEP monitoring in the clinical setting

In surgery, SEPs find use in many situations such as:

1. Operative retraction and manipulation around the spinal cord including clamping of the abdominal aorta during operative correction of an aneurysm.
2. Orthopedic cases where the spinal cord may be at risk such as Harrington rod insertion and removal.
3. Carotid endarterectomy where clamping of the artery may ensue cerebral hypoxia.
4. Cerebral aneurysm clipping, where the clip may endanger sufficient blood flow to other parts of the artery.
5. Deliberate hypotension during cardiopulmonary by-pass to avoid episodes of malperfusion of the brain.
6. For the evaluation of peripheral nerve lesions.
7. For the determination of sufficient analgesia intraoperatively.

Application of SEP monitoring in the ICU is another task where central nervous system functions can be evaluated. With the various physiological parameters monitored continuously at the bedside, there is practically no monitor which gives serial or continuous assessment of neuronal function. Especially SEPs in such a situation present a valuable tool to:

1. Determine the functional deficit in head injured patients when muscle paralysis or high-dose barbiturate therapy has made physiological examination unreliable.
2. Monitor the brain injured patient in the acute phase during which the dynamic condition of the brain (ischemia, cerebral blood flow, intracerebral pressure, cerebral edema) affects survival of brain function.
3. Assess critical conditions by medical or surgical management enhanced by serial monitoring of brain function.
4. Assess coma during long-term sedation when the patient is ventilated.
5. Differentiate between peripheral and central dysfunction.
6. Evaluate nerval root lesions.
7. Evaluate dysfunction of the cervical spine.

13.2 Representative examples of SEP traces at different clinical situations

The following section will reflect on some SEP traces focusing on the indications where SEP monitoring has aided in diagnostic finding intraoperatively and in critical care therapy. Most of the patients were monitored with the Lifescan™ Evoked Potential Monitoring System.

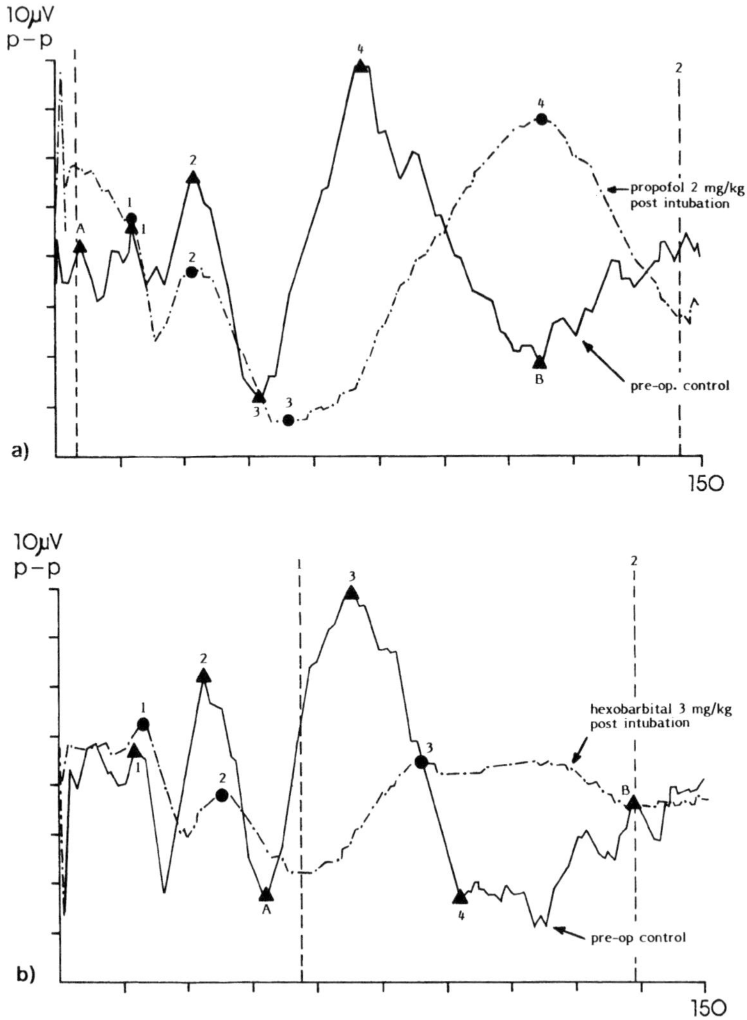

Fig. 87. Representative example of two agents used for the induction of anesthesia and their corresponding effects on median nerve somatosensory evoked potential changes after laryngoscopy (time axis in ms).
(a) after propofol 2 mg/kg.
(b) after hexobarbital 5 mg/kg.

The induction period. Induction with hexobarbital will result in a marked reduction of amplitude of the early but more so of the late (over 50 ms) peak when compared to the preoperative control period (Fig. 87). This effect on the SEP suggests sufficient inhibition of nociceptive impulses which may evolve following laryngoscopy for the insertion of the endotracheal tube.

In comparison with another induction agent (propofol) which has a pronounced hypnotic component, amplitude suppression is not that intense. The following intubation manoeuver is not sufficiently blocked to prevent afferent volleys from inducing sympathetic discharge. This effect of propofol is due to a paucity of analgesic activity and antitussive action; however it induces a sufficient depth of sleep.

Intraoperative monitoring. During the anesthetic procedure SEP can be helpful in determining the depth of analgesia. Although the patient may be sufficiently asleep, with the aid of SEP monitoring an insufficient blockade of nociceptive impulses can be detected. This reaction is foreseen earlier than through some of the cardiovascular parameters usually used in determining a sufficient anesthetic depth; the latter have a lag of up to several minutes.

A typical recording of a female (age 68) undergoing gall bladder removal is shown in the next figure (Fig. 88). Although the patient is sufficiently deeply anesthetized with propofol (0.1 mg/kg/min) -N_2O/O_2-relaxant anesthesia, the SEP reflects an increase in amplitude and a reduction of latency during the time where the surgeon pulls at the liver.

The peaks of the SEP are similar to the preinduction period suggesting insufficient analgesia. It was only after 3 minutes that blood pressure and heart rate, other indicators of insufficient anesthesia increased. This was corrected by the additional injection of the opioid fentanyl.

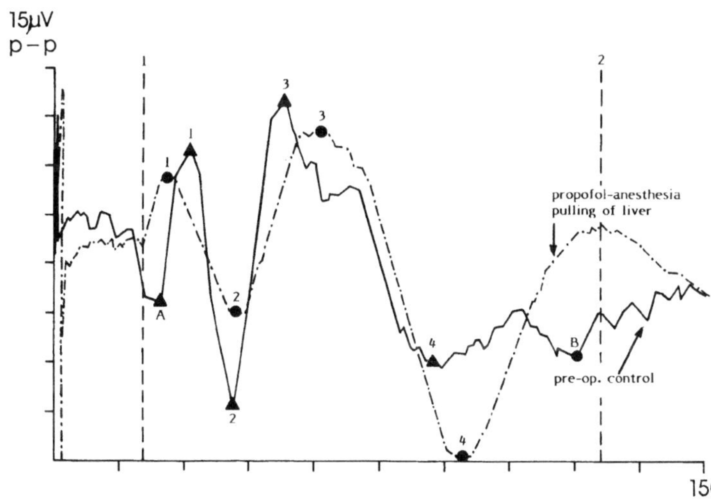

Fig. 88. Intraoperative monitoring during propofol-nitrous/oxide-oxygen anesthesia with traction at the mesenterium results in an increase in amplitude and a reduction of latency of the SEP reaching preanesthetic level (time axis in ms).

Sufficient blockade of nociceptive impulses during the surgical manipulation is represented in another example. Compared to the control-preoperative period (Fig. 89a), anesthesia with enflurane N_2O/O_2 at 1.5 vol% in a young male induces a suppression of the late (> 50 ms) evoked peak. This suppression, however,

Fig. 89(opposite page). Intraoperative monitoring of sufficient blockade of afferent nociceptive input using somatosensory evoked potentials.
Compared to control (a) an increase in surgical stimuli results in an increase in amplitude height of the late peak (b). The insufficient analgesia is attenuated by an increase in the concentration (vol%) of the anesthetic enflurane (c). x-axis = ms/division = 15; y-axis = μV/division = 1,25.

Patient (*14.05.1965) polytoxicomania and tibia bone fracture

		L	A	L	A
a)	Peak 1:	101.22 03.136		026.88	−03.648
	Peak 2:	056.75	−04.352	037.17	01.664
	Peak 3:	036.17	00.512	057.41	−02.624
	Peak 4:	024.56	−03.584	101.22	03.008
b)	Peak 1:	026.55	−01.600	027,54	−02.432
	Peak 2:	041.15	01.536	039.83	01.856
	Peak 3:	062.06	−00.032	078.32	−04.224
	Peak 4:	087.95	01.248	131.43	05.408
c)	Peak 1:	027.21	−01.568	027.54	−02.432
	Peak 2:	039.49	01.376	039.82	01.856
	Peak 3:	064.05	−01.728	078.32	−04.224
	Peak 4:	094.59	02.240	131.43	05.408

L = Latency (ms); A = Amplitude (μV).

is overridden when surgical manipulation induces an increase in amplitude height. (Fig. 89b). Followed by an increase in concentration of the volatile anesthetic to 2 Vol%, it is only after 5 min (when sufficient brain concentrations are reached) that the 'excited' afferent somatosensory pathway is sufficiently inhibited in regard to nociceptive input and the generation of evoked potentials (Fig. 89c). Thus, the SEP not only is useful to evaluate analgesia, as in experimental human algesimetry, but the SEP also has the potential to give advance warning of insufficient analgesia during the operative procedure [84].

13.3 Postoperative use of SEPs

Evoked potentials monitored in a patient in the immediate postoperative period can be used for comparison with the preinduction period to evaluate whether the patient has sufficiently regained all his reflexes and there is no overhang of the anesthetic. But more so, SEPs are useful to give an idea whether the impaired neurophysiological response suggests intraoperative brain damage.

In Fig. 90, the evoked potential, 20 minutes after the end of anesthesia, has a similar appearance as during the preoperative control period. Only the early peak (number 2) still is depressed in regard to amplitude height, reflecting a residual overhang of the anesthetic (enflurane N_2O/O_2-relaxant anesthesia).

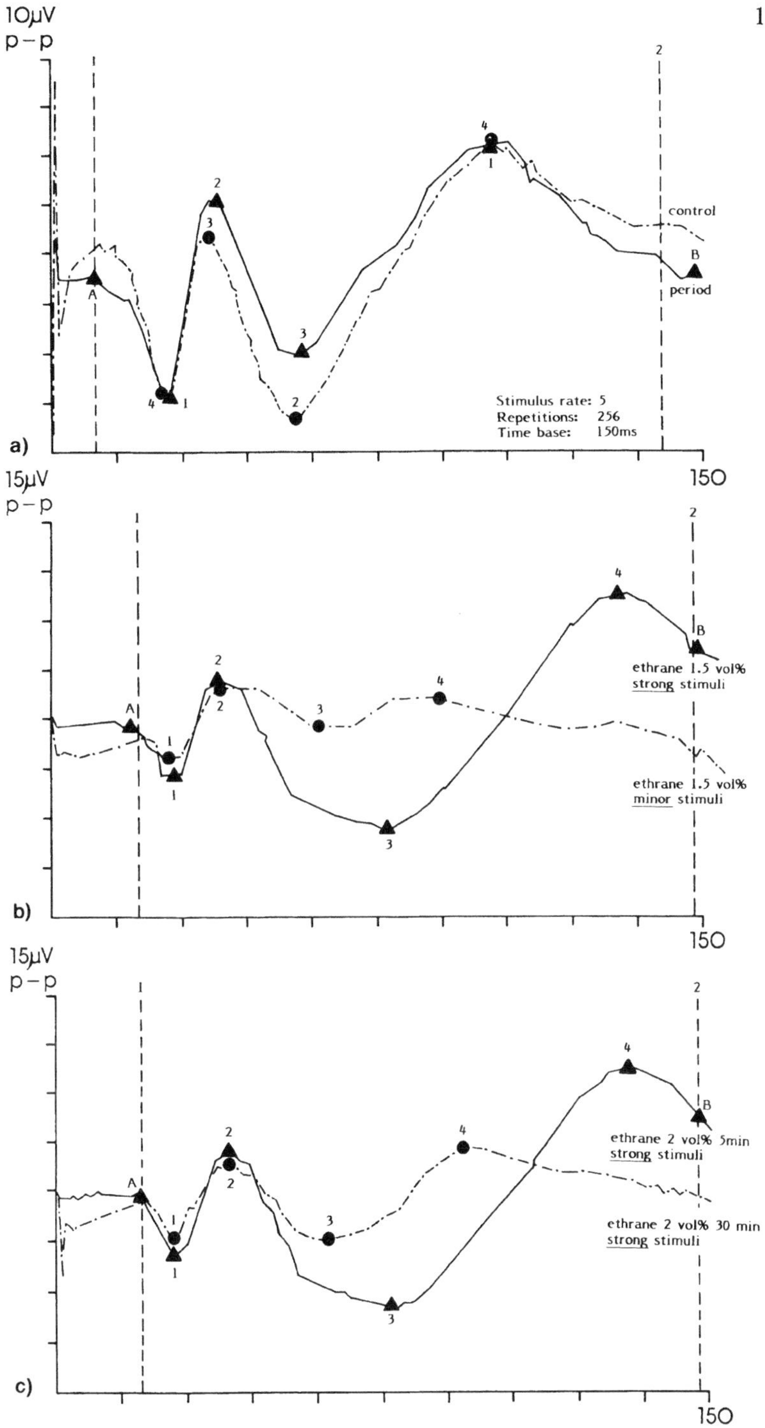

Fig. 89. Intraoperative monitoring of sufficient blockade of afferent nociceptive input using somatosensory evoked potentials.

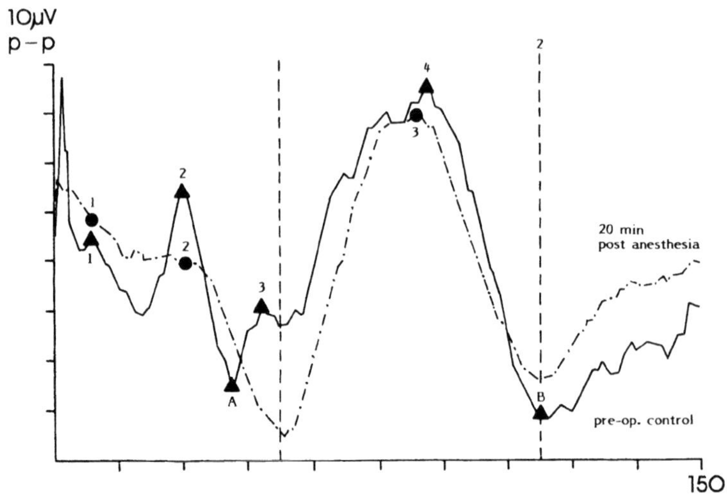

Fig. 90. Postoperative evaluation of a possible overhang after propofol N_2O/O_2-anesthesia using evoked potential monitoring. *Note*: 20 minutes post anesthesia the peaks are practically identical to the ones taken preoperatively (time axis in ms).

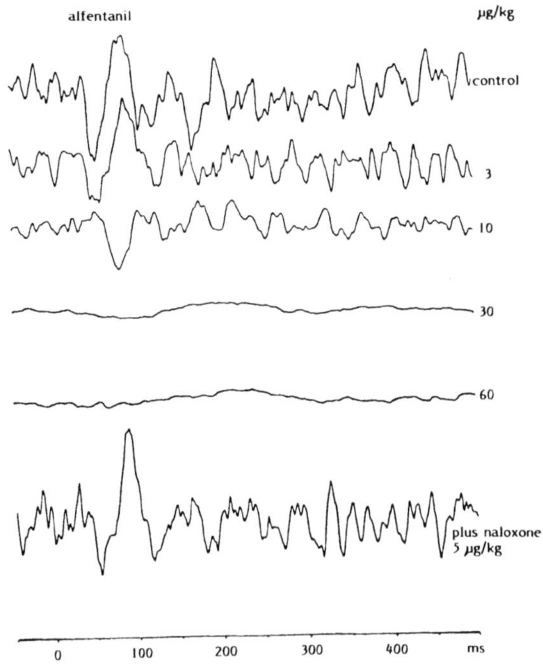

Fig. 91. Somatosensory evoked potentials and increasing doses of the opioid alfentanil. Due to total receptor occupation loss of the cortical response will result. This effect is reversed instantly by competitive displacement of the opioid by naloxone, a specific antagonist.

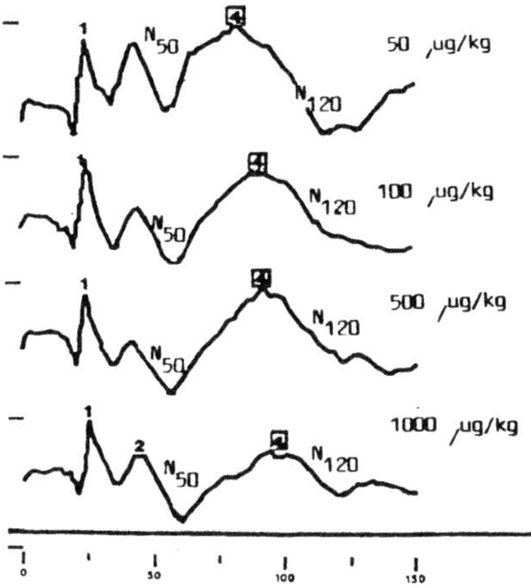

PEAK	LATENCY mS	AMPLITUDE µV
1	23.43	+ 1.328
2	42.18	+ 0.781
3	53.90	- 1.328
4	82.03	+ 1.484
N_{120}	116.01	- 2.578
1	23.43	+ 1.953
2	44.53	+ 0.078
3	58.59	- 2.343
4	87.89	+ 1.875
N_{120}	121.87	- 1.015
1	23.43	+ 2.109
2	42.18	- 0.078
3	56.25	- 2.500
4	90.23	+ 1.953
N_{120}	119.53	- 0.703
1	23.43	+ 2.500
2	43.35	+ 0.859
3	59.76	- 1.953
4	96.09	+ 1.015
N_{120}	120.70	- 0.703

Fig. 92. The use of somatosensory evoked potentials in the evaluation of the analgesic effectiveness of nalbuphine, a mixed narcotic analgesic. Higher doses than 500 µg/kg result in a 'ceiling effect', i.e. the late N_{120} peak cannot be depressed any further.

As mentioned earlier, the use of high-dose opioid anesthesia may result in a total loss of early as well as late evoked potentials. Such a situation, otherwise considered as a neurological deficit due to ischemia, is related to the anesthetic which blocks all afferent volleys from crossing the synaptic cleft at the nerve junction (Fig. 91). Potent opioids thus inhibit nerve impulse afferents by selective blockade of receptor sites at the synapse. The following postjunctional deficit resembles the picture of a complete functional loss. This effect is related to the number of receptors in the somatosensory tract being occupied and can be completely reversed by the specific antagonist naloxone [82].

Also, the net effect of mixed narcotic analgesics such as nalbuphine can be evaluated in the human, taking the SEP as a method to determine analgesic effectiveness. In Fig. 92 increasing doses of the analgesic result in a dose-related decrease of amplitude of the late (N_{120}) peak. It can be seen that at 500 µg/kg a ceiling effect in regard to amplitude reduction is reached, as doubling of the dose leads to no further decrease of amplitude height. Thus, evoked potentials can be of use for testing in algesiometry when new compounds are evaluated in regard to their effectiveness and compared with other existing drugs. A similar use of evoked potentials has been described by other authors who claim that late evoked potentials (over 100 ms) correlate closely with the subject's individual feeling of pain sensation *but not* with stimulus intensities [84–86].

The peaks with latencies between 30–70 ms (lower extremity) or between 13–30 ms (upper extremity) appear to be most stable for monitoring purposes during operation, as they are changed the least by volatile anesthetics. For monitoring of head injury or operations where the cerebral blood flow is at risk, such as during carotid endarterectomy or superficial temporal artery to middle cerebral artery bypass, the use of the central conduction time (CCT) from the cervical response (usually C_2) to the sensory cortex is of considerable usefulness.

14. Use of evoked potentials in the ICU

14.1 Introduction

In the ICU, serial analysis of cortical evoked potentials will give an idea of the possible status in patients with head trauma. A representative example shows a patient with severe head trauma who had been involved in a car accident. The SEP shows a marked difference in regard to right and left median nerve stimulation (Fig. 93a). Although CT scan revealed no signs of intracerebral mass bleeding, there was a left arm paralysis, a left sided mydriasis, with a radiologically verified fracture at the base of the scull. The pathologically reduced amplitude of the early potentials during left median nerve stimulation suggested a left-sided functional deficit involving the medulla. In the following days of intensive care therapy, sings of recovery became evident as the amplitude increased in height being similar to right median nerve stimulation (Fig. 93b-c).

Thus, the evoked potential was not only able to detect minute lesions in subcortical structures which otherwise are undetectable by CT scan, it also serves as a prognostic tool to evaluate the efficiency of therapy in the long run.

Somatosensory evoked potentials are also helpful when it comes to diagnose a potential peripheral nerve lesion after polytrauma.

In a young female after a car accident with a fracture of the right forearm and head trauma, operative exploration of the peripheral nerve for decompression was considered necessary in order to avoid nerve lesion. While the comatose patient was on the ventilator and the right arm was in a cast, right median nerve stimulation revealed functional integrity with a sufficient evoked potential at Erb's point and no depressed amplitude of cortical evoked potentials (Fig. 94). Thus, an otherwise unnecessary operation for the patient could be avoided.

In the intensive care evoked potentials can give the clinician more information about the patient's central nervous function than any other monitoring system.

Fig. 93 (opposite page). Example of a comatose young caucasian male after severe head trauma where median nerve evoked potentials reflect side difference. During therapy in the ICU the suppressed amplitude at left median nerve stimulation recovers, suggesting functional improvement. The latter was also reflected in the patient's general behavior as he regaind consciousness. x-axis = ms/division = 5; y-axis = μV/division = 1,25.

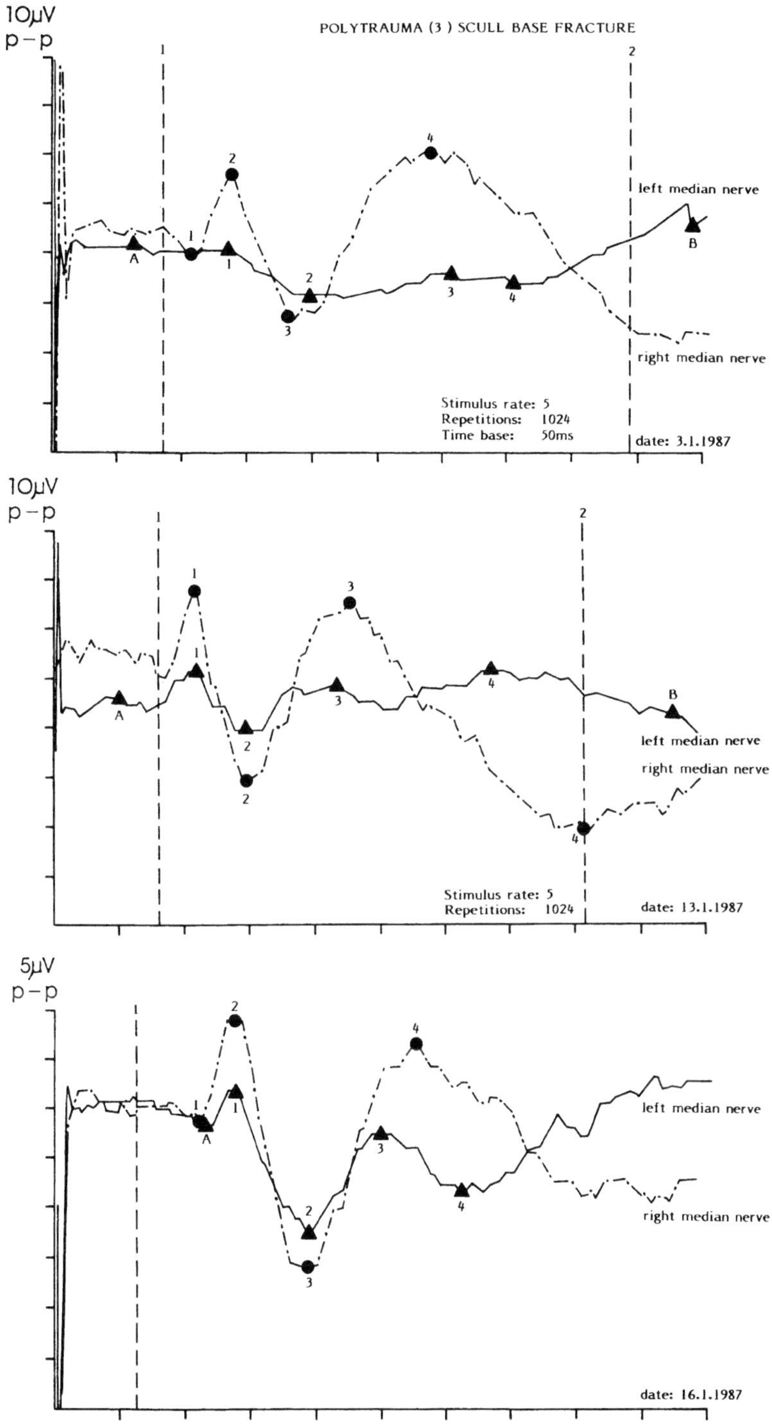
10µV
p-p
POLYTRAUMA (3) SCULL BASE FRACTURE
left median nerve
right median nerve
Stimulus rate: 5
Repetitions: 1024
Time base: 50ms
date: 3.1.1987
10µV
p-p
left median nerve
right median nerve
Stimulus rate: 5
Repetitions: 1024
date: 13.1.1987
5µV
p-p
left median nerve
right median nerve
date: 16.1.1987

Fig. 93.

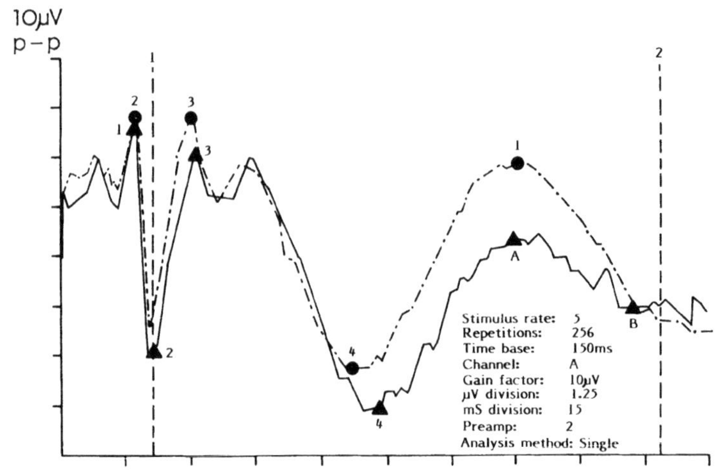

Sedation with fentanyl 2 ml/hour and respirator therapy (IMV).

	L	A	L	A
Cursor 1:	021.24	–01.856	108.19	00.576
Cursor 2:	144.04	–01.792	136.07	–01.408
Delta:	122.80	00.064	027.88	–01.984
Peak 1:	108.53	02.688	017.59	03.552
Peak 2:	017.92	03.936	021.90	–02.752
Peak 3:	030.53	03.680	031.53	02.752
Peak 4:	070.69	–03.232	075.00	–04.224

L = Latency (ms); A = Amplitude (µV).

Fig. 94. Right median nerve evoked potentials derived in a comatose young female with questionable peripheral right arm nerve compression. The normal SEP suggests functional integrity.

A 65 year old patient who had undergone aneurysm operation of the thoracic aorta, developed thrombosis of the vena cava, and meningitis arising from the ear postoperatively. He was in deep coma and an appalic syndrome was suspected. The EEG power spectra showed pathological slow delta to theta wave activity over both hemispheres. Median nerve evoked potential corroborated the diagnosis of cortical brain damage as only the early peak was derived while the late peak was practically flat. This suggested functional deficiency of cortical associative fields as the patient had not received any sedatives or opioids over the past days (Fig. 95).

Thus, the SEP is not only helpful in substantiating a suspected diagnosis, it can also be used to evaluate the effectivity of therapy and the possible prognostic outcome.

During CABG (coronary aortic by-pass grafting) hypothermia has a significant effect not only on peak latency changes, but also on the amplitude of the late peaks. This is reflected in an example of a 55 year old male undergoing by-pass grafting. In comparison to normothermia with enflurane-N_2O/O_2 and fentanyl (Fig. 96a), deep hypophermia (22 °C) induced an increase in latency and a marked suppression of amplitude (Fig. 96b-c). This effect slowly reversed on rewarming. In the post-by-pass phase, normal temperatures resulted in similar evoked potentials as in

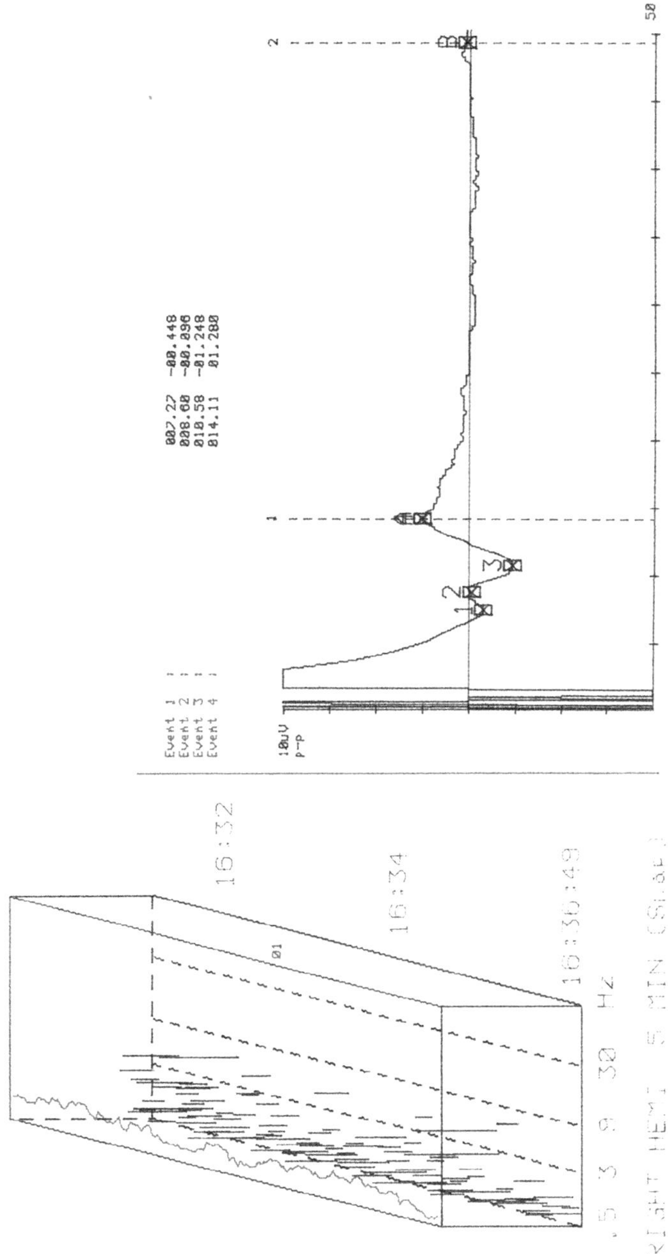

Fig. 95. Posttraumatic deep coma in a patient with pathological delta activity in the EEG power spectra. In the SEP only the early peak can be derived suggesting cortical lesion.

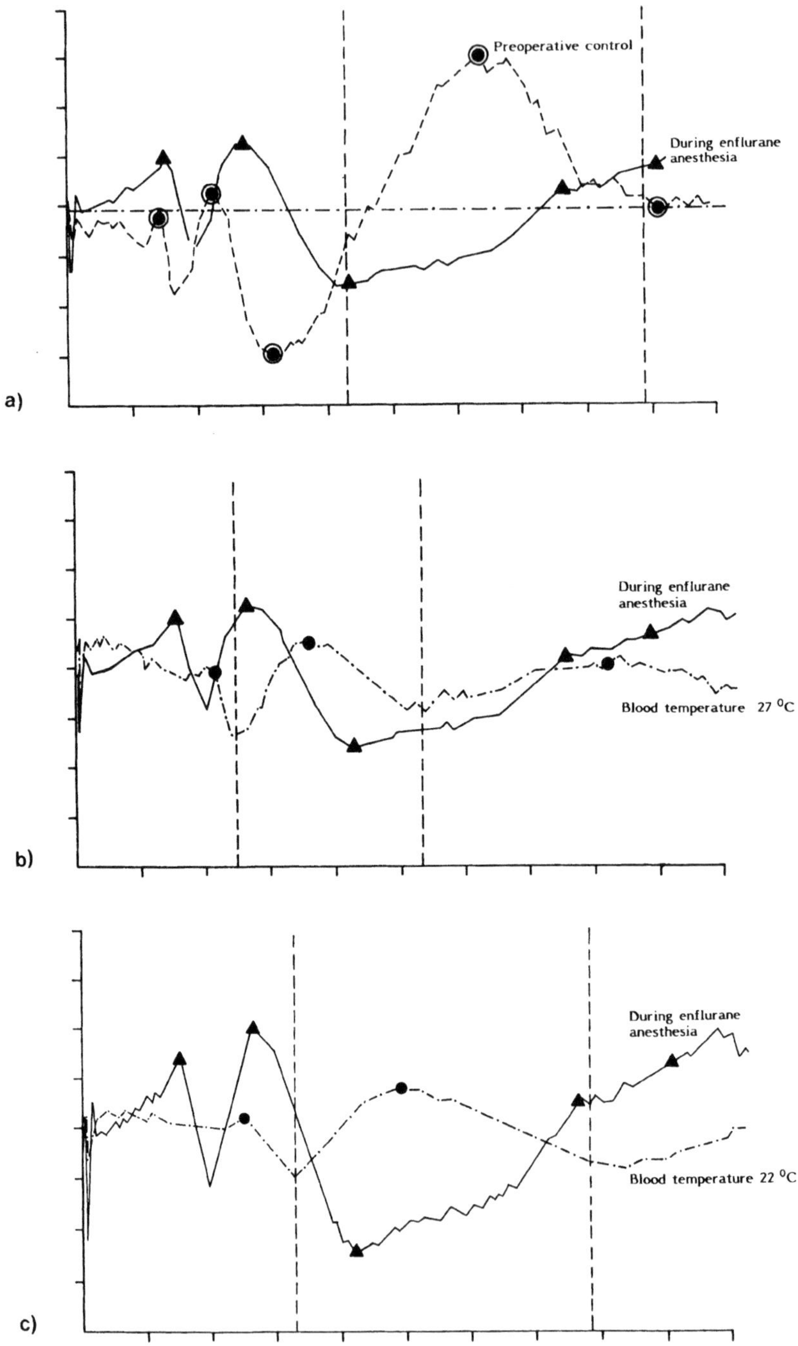
Preoperative control
During enflurane anesthesia
a)
During enflurane anesthesia
Blood temperature 27 °C
b)
During enflurane anesthesia
Blood temperature 22 °C
c)

Fig. 96a–c.

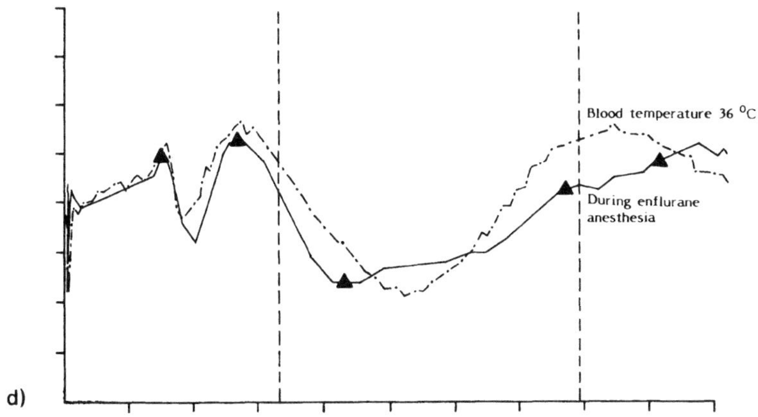

Fig. 96. The effect of cooling on median nerve SEPs in a patient undergoing CABG. Hypothermia induces latency changes and amplitude depression similar to the effects of malperfusion.
(a) Post induction of anesthesia followed by ventilation with nitrous oxide/oxygen (2:2) plus enflurane (1.5 vol%).
While being on by-pass, hypothermia is induced and blood temperature drops to 27 °C (b). This is followed by a further decline to 22 °C (c). After rewarming to a blood temperature of 36 °C there is a recovery of evoked potentials (d).

the pre-by-pass phase (Fig. 96d). Thus, hypothermia induces effects on the SEP which are similar to cerebral malperfusion. Therefore, it is always advisable to look at other parameters before making a final statement in regard to SEP changes. As demonstrated in Fig. 96a, fentanyl in a dose of 0.35 mg had practically little effect on the early (< 50 ms) evoked peak, which corroborates the results of other research groups [86]. Depression of CNS function during hypothermia, however, is reflected in the central conduction time (CCT). It shows the time lag of the evoked peak derived at C_2 to the cortical peak N_{20}. Cooling induces a synaptic delay of axonal conduction between wrist and cortex. During stable hypothermia, the CCT thus is a useful guide to monitor the cerebral status of the patient.

14.2 SEPs in the diagnosis of lesions in the plexus of the upper extremities and in cervical root lesions

Besides evaluating central nervous functional pathways, the SEPs can be helpful in the diagnosis of traumatic root lesion, for example of the brachial plexus. In such an incidence the functional deficit is in all parts of the corresponding nerve site while the contralateral side shows normals EPs. Thus, not only is the cortical EP flat, but also at Erb's point and at C_7 no evoked potentials can be derived. A functional deficit in the area of the cervical root, i.e. rostral to the spinal ganglia, however, results in no change of amplitude and latency of the evoked potential recorded over the ipsilateral Erb's point. In those cases where the motoric axons show partial degeneration, a slight amplitude reduction will be observed as the EP

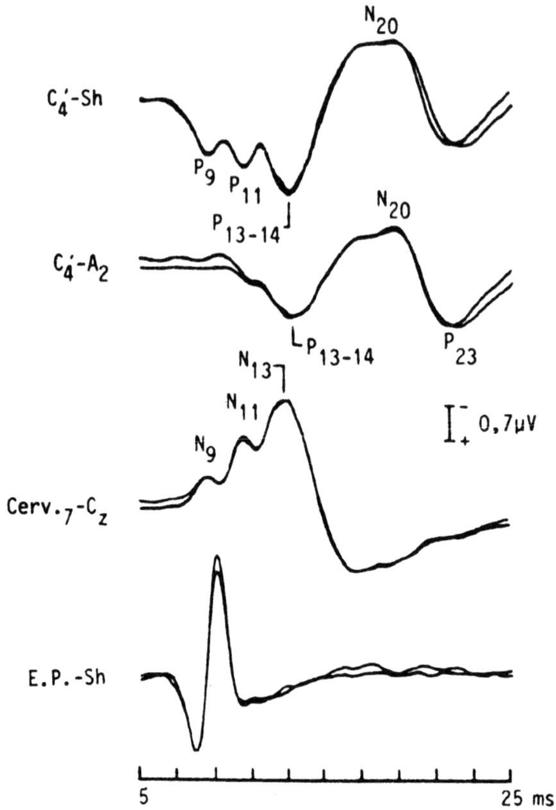

Fig. 97. Short latency evoked potentials during median nerve stimulation using different reference sites. EP = Erb's point; Sh = Shoulder; A_2 = Right ear lobe.

from Erb's point arises from mixed nerves in which the motoric axons partly generate the potential. If, however, the nervous pathways between Erb's point and cervical spine is affected in one way or another (tumor infiltration, lesions of the cervical root by means of trauma, or local infection), peaks N_{11} and N_{13} are either reduced in amplitude or they are completely absent. These latter peaks are best visualized when the usual contralaterally placed recording electrode has a contralateral shoulder as reference, or when the recording electrode over C_7, C_z is taken as reference (Fig. 97). In such instances, the potential at Erb's point has no deficit as the point of obstruction is proximal to Erb's. According to standard thinking N_9, N_{11} and N_{13} are generated in the brachial plexus, the posterior root of the spine, and the ascending afferent nervous tract respectively, while N_{14} is generated in the lemniscal tract (Fig. 98) [88–90]. However, it has to be underlined that all measurements have to be done on both sides in order to compare amplitude and latency changes. The N_{13} peak, as it is generated in the lower brain stem, is best visualized when the recording electrode is placed over the second spinal vertebra C_2 (Fig. 98).

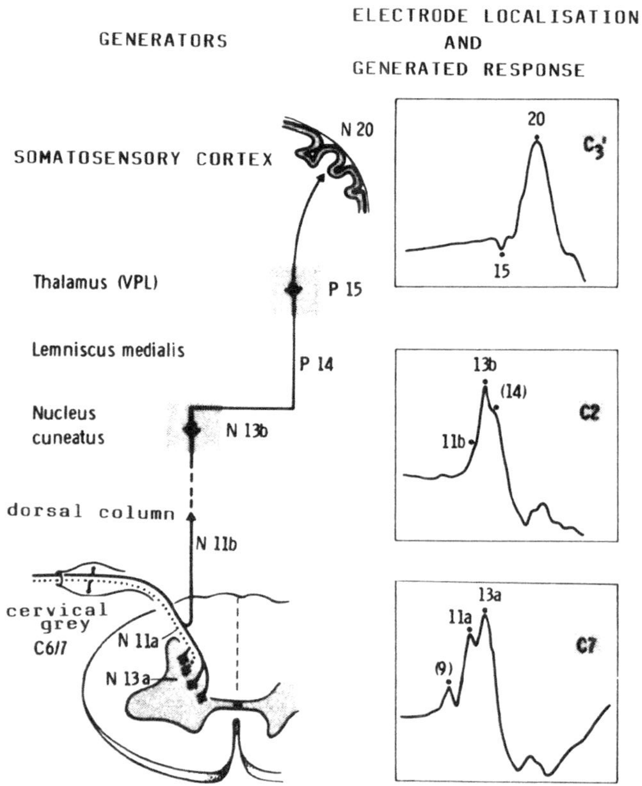

Fig. 98. The generators of spinal and cortical SEPs during median nerve stimulation.

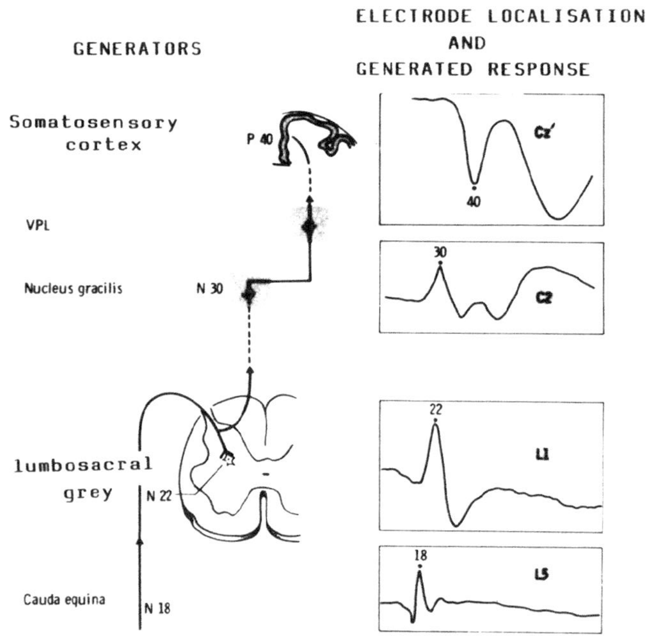

Fig. 99. The generators of spinal and cortical SEPs during posterior tibial nerve stimulation.

Table 5. Normative values during median nerve stimulation.

Latency (ms)	Erb	C_7	C_2	N_{20}
Mean value	10.2	13.5	13.7	19.3
± SD	± 0.9	± 0.9	± 0.9	± 1.2
Upper limit (mean ± 2.5 SD)	12.4	15.8	15.9	22.3
Maximum right-left difference	0.7	0.7	0.7	1.1
Latency interval (ms)	**Erb/C_2**	**Erb/C_3'; C_4'**	**C_7/C_2**	**C_2'; C_3'–C_4'**
Mean interval	3.4	9.0	0.2	5.8
± SD	± 0.6	± 0.8	± 0.2	± 0.6
Maximum interval (mean ± 2.5 SD)	4.9	11.0	0.6	7.3
Maximum right-left difference	0.6	0.7	0.6	1.0
Amplitude (μV) base to peak				
Mean value	3.7	1.6	1.6	2.3
± SD	± 2.3	± 0.7	± 0.7	± 1.0
Range	0.8–12.3	0.4–4.1	0.4–3.9	0.6–5.3
Side difference (%)	48	36	38	46
Amplitude quotient	**Erb/C_7**	**C_2/C_7**	**C_3'; C_4'/C_7**	
Mean	2.3	0.98	1.4	
Range	1.1–8.8	0.7–1.7	0.7–8.9	

Adapted from Stöhr [96].

Additionally, the electrode at C_2 is useful in determining the CCT (central conduction time) when the contralateral cortical evoked potential is recorded with a scalp electrode. The N_{13}/N_{20} transit time as well as the N_{20}/N_{13} amplitude quotient are also useful in the diagnosis of deficit in transit between the cervical spine and the primary sensory cortex. A CCT above 7.25 ms and a quotient below 0.65 μV can be considered pathological (Tables 5–6).

From the above it is conceivable that a diagnosis can be derived using various quotients. Thus, any reduction in the C_2/C_7 value is an index for a functional deficit rostral to the cervical spine C_7. The other quotient N_{20}/C_2 reflects important information in regard to a partial deficit between the cervical spine and the primary sensory cortex, especially when the value drops below 0.65 μV.

Table 6. Normative values during posterior tibial nerve stimulation.

Latency (ms)	N_{18}	N_{22}	N_{30}	P_{40}
Mean value	18.4	21.7	29.5	38.8
± SD	± 1.2	± 1.6	± 1.9	± 2.0
Upper limit (mean ± 2.5 SD)	21.4	15.8	34.3	43.9
Latency interval (ms)	N_{18}/N_{22}	N_{22}/N_{30}	N_{30}/P_{40}	N_{22}/P_{40}
Mean interval	3.8	7.9	8.8	17
± SD	± 0.9	± 1.0	± 1.6	± 1.7
Maximum interval (mean ± 2.5 SD)	6.0	10.4	12.9	21.3
Maximum right-left difference	2.66	2.41	2.81	3.5
Amplitude (μV) base to peak				
Mean value	0.3	0.6	0.6	1.8
± SD	± 0.2	± 0.3	± 0.3	± 1.3
Range	0.05–0.8	0.15–1.1	0.15–1.25	0.35–5.2
Maximum side difference (mean ± 2.5 SD)	0.51	0.5	0.54	2.5
Amplitude quotient	N_{22}/N_{18}		P_{40}/N_{22}	
Mean	3		4.9	
Range	1.1–6.6		0.85–27.3	

Adapted from Riffel [99].

Involvement of the medulla by trauma, degeneration or by infiltration of a tumor, results in changes of the SEP with the recording (active) electrode over C_2 during median nerve stimulation. In such an instance the usually well definable N_{13} peak is deformed while cortical recording depends on the extent of the affection, followed by a deformation of amplitude and reduction as well as an increase in latency.

The *locked-in-syndrome* which is due to an infarct in the pontine area results in a total deficit of voluntary movements with the exception of vertical eye movements. The SEP in such a case is characterised by a total loss of early as well as late cortical EPs with a N_{13} peak clearly visible over the cervical spine C_2. Any reduction in the amplitude by at least 40%, when compared to the other side, is a clear index of

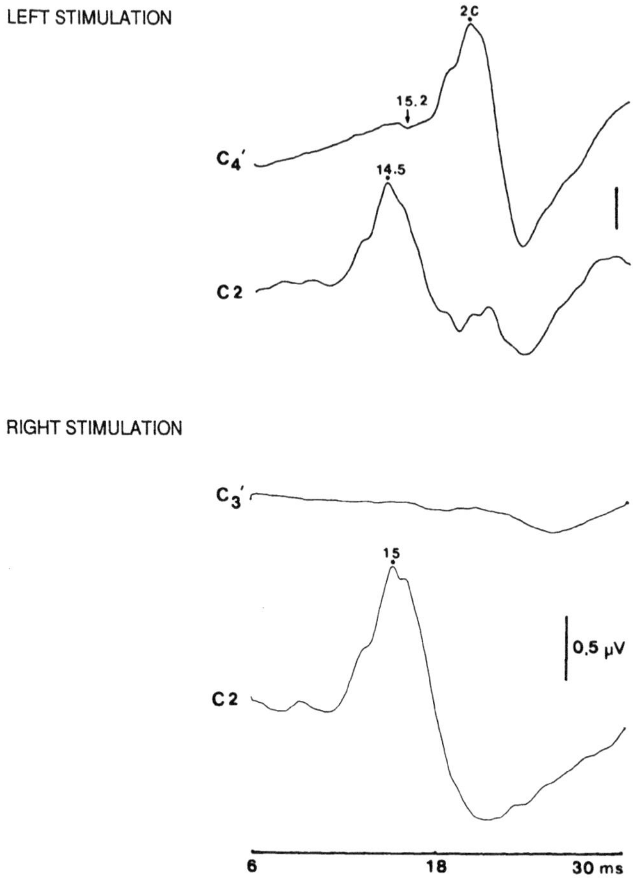

Fig. 100. Left-sided intrathalamic bleeding and the corresponding median SEP (adapted from Stöhr [100]).

pathology which is characteristic in the locked-in-syndrome, as well as in post traumatic deficits due to bleeding in the thalamus (Fig. 100).

If the infarcted or traumatized area is localized posterior to the postcentral gyrus a normal N_{20} peak, the primary complex, will result even though the late peaks are completely missing. Such a picture can be interpreted as presence of functional integrity of the postcentral gyrus. Any lesion that extends from the cortex down to the internal capsula will result in a total loss of all cortical components. In certain cases it is advisable not to take FpZ as the point of reference, but to choose instead the ipsilateral shoulder as reference and C_3' or C_4' as the active electrode. Once FpZ and C_3' have similar negative polarities, it is conceivable that in an electrode montage FpZ-C_3' the resultant potential will be zero (Fig. 101).

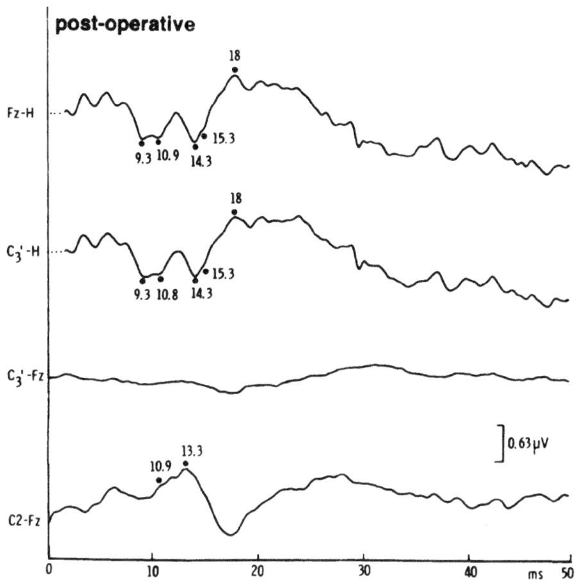

Fig. 101. SEP recording after a neurosurgical operation. At C_3'–FpZ position no potential can be derived. If, however, the hand or the shoulder is taken as reference, a peak can be derived. This is due to the similar polarities at C_3' and FpZ.

14.3 Use of SEP monitoring in head trauma, vascular disease, and brain death

Of clinical use in the intensive care unit are SEPs when a barbiturate therapy in traumatic coma, used for the protection of the brain blunts the usual clinical methods of diagnosis.

In addition to the usual indications of SEP measurements in head trauma, this diagnostic tool is of prognostic value. Besides measuring the cervical (C_7, C_2) and cortical evoked potentials during median nerve stimulation, the interpeak latencies C_2/N_{20}, also referred to as the central conduction time (CCT), as well as the interrelation of the amplitudes N_{20}/N_{13}, can be useful parameters. Similar to the usual neurologic evaluation, the non-traumatized side serves as comparison. A normal or slightly prolonged CCT and a normative or only slightly diminished ratio amplitude reflects a good prognosis.

A marked prolongation of CCT, however, and especially an amplitude reduction of cortical responses in addition to the low amplitude quotient N_{20}/N_{13} strongly suggest a poor outcome. Any bilateral loss of the cortical response can be considered of bad prognostic value.

In the diagnosis of cerebral death, the SEPs, even in the case of bilateral loss of cortical responses, up till now are not considered to be a definite sign of permanent brain death. Such a diagnosis can be reached with the aid of early AEPs. This is deduced from the fact that any irreversible loss of brain stem activity is a prerequi-

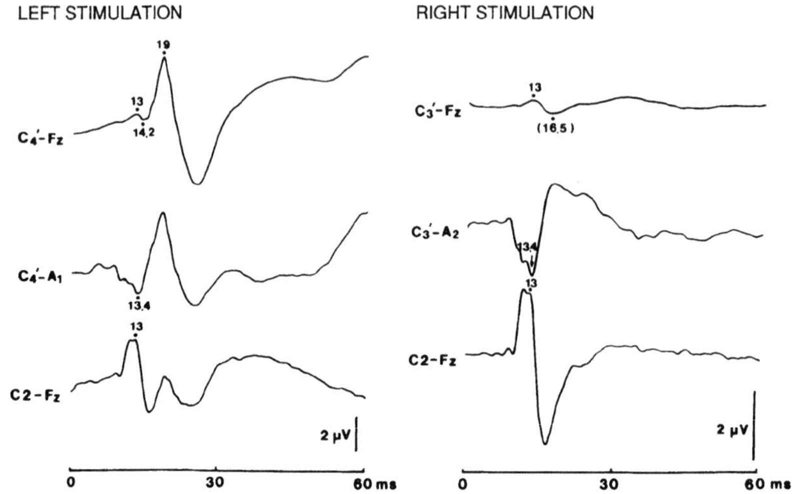

Fig. 102. CT scan verified infarct of the cerebral media artery and its effect on the median SEP. The late peaks over the left hemisphere are missing.

site of cerebral death. However, in addition to the bilateral loss of cortical evoked potentials during SEP measurements the peak derived from C_2 (the N_{13} peak) can be considered a useful sign of an intact medulla as this peak is generated in the lower part of the brain stem. Brain death can thus also be *diagnosed* with the *SEP*, especially when the active electrode over C_2 detects no peak. Any primary loss of that potential after head injury is a sign for a combination of cortical and lower brain stem involvement.

In vascular attacks such as an infarct of the middle cerebral artery the early phase in the SEP is characterized by a marked distorted shape with amplitude suppression which is related to the area of hypoxemia [91]. SEP diagnosis in this regard can be used as an aid to control the course of the ailment (Fig. 102). Also, in transitory ischemic attacks the increase of latency of the SEP can be used to monitor the functional deficit of the underperfused area [92].

Similar to cortical evoked potentials, being a diagnostic tool during cerebral hypoxemia, hypoxic phases of the spinal cord may also be detected, as they appear for instance during aortic cross-clamping of an abdominal aneurysm. Spinal cord ischemia very soon after aortic clamping is characterized by a marked depression of amplitude which finally may result in a total loss of any signal ascending to the cortical level. As different protective techniques are fostered to increase survival time of the spinal cord, SEPs are used as a tool to evaluate any residual hypoxic effect after declamping [93]. An alternative way to monitor spinal cord function during aortic surgery is the recording of the evoked potential from the scalp produced by *epidural* stimulation at L_3/L_4. Spinal cord potential disappearance with *tibialis* stimulation often is not caused by ischemia of the lumbar spinal cord, but by ischemia of the peripheral nerve. Such monitoring during aortic cross-clamping is the only way of early detection of ischemia which otherwise is masked by anesthesia.

From the above and in accord with the literature [94–95], the usefulness of SEP monitoring during vascular and open heart surgery can be summarized as follows:

1. Any increase of CCT above 20% with a depression of amplitude of the N_{14} and N_{20} peak during carotid clamping which after declamping will not resolve, is bound to end with a neurological deficit.
2. Any drop of mean pressure during ECC (extracorporeal circulation), accompanied by a reduction of amplitude in the order of 20–50%, is a monitoring sign of insufficient perfusion and should be followed by an increase of perfusion pressure.
3. In aneurysm clipping of the common cerebral artery, the SEPs derived during posterial tibial nerve stimulations are likely to pick up any deterioration in blood distribution. Specific changes of the early N_{20} or P_{25} peak can be seen. Thus, a time axis of 50 ms is sufficient for evaluation. Clipping of the so-called Pica aneurysms should be done under continuous monitoring of SEPs in order to evaluate border-line perfusion and pick up insufficient blood supply to cerebral areas.
4. In aneurysm clipping of the anterior cerebral artery, malperfusion is best diagnosed during posterior tibial nerve stimulation. Any increase in CCT signalling malperfusion in this pathology, cannot be considered a reliable sign.
5. Hypoxic effects during aneurysma clipping of the posterior cerebral artery are monitored best when the active electrode is positioned at P_3 or P_4 rather than at position C_Z.

In any case, with the aid of SEPs, the cerebrovascular reserve can be monitored directly during the clipping procedure *before* permanent damage will evolve. In spite of the potential advantages SEP monitoring in the OR and/or the ICU may bring, it should be remembered that the evoked potential is only a warning signal of malfunction of the specific pathway in the nervous system. Potential drawbacks of this methodology are:

1. It takes time for the averaging procedure, depending on the number of sweeps (sometimes up to 3 minutes).
2. Only the *afferent* pathway in regard to its function is monitored, not the *efferent* pathway.
3. The SEP is not a *diagnostic* instrument to tell the physician about the cause of the underlying malfunction.
4. Neuromonitoring after head trauma will only be able to diagnose a functional deficit. Usually there is a good correlation of CT and SEP. In some cases, however, the SEP is a better diagnostic system, as the former will only evaluate changes in morphology. If density difference is below 8 mm, the CT is unable to visualize these changes.
5. Normal SEPs provide no direct evidence of intact motor function as the major parts of the CNS tested during SEP monitoring are the dorsal columns of the spinal cord. In acute spinal trauma and during operations involving the

spine, fortunately sensory and motor function usually correspond closely [96].

6. Irreversible carotid obliteration can be associated with postoperative SEP deficit in the monitoring pathway. However, when obliteration is reversed, the resulting waveforms will return to normal before the end of anesthesia signaling a preservation of function.
7. The central conduction time (CCT) of SEPs has a close correlation to the prognostic outcome in comatose patients.

14.4 Pitfalls and pointers for SEP measurement in the OR and the ICU

Anesthesia: Avoid inhalation agents when monitoring intermediate or long-latency evoked potentials. Obtain pre-anesthesia base-line recording. Avoid bolus injection of CNS active drugs during monitoring periods.

Circulation: Monitor intraarterial pressure. Avoid fluctuation of perfusion pressure during critical phases of measurement.

Electrodes: Use silver/silver chloride or gold electrodes. Attach electrodes with collodion. Protect electrodes with tape or gauze. If used within the surgical field, platinum subdermal electrodes should be used.

Muscle relaxants: Somatosensory stimulus should be supra-threshold prior to the administration of a muscle relaxant.

Noisy recording: Electrode impedance should be less than 3 kOhms and matched. If in doubt inspect the raw EEG for artifacts. Protect the electrodes under surgical drapes. Twist electrode lead wires together. Suspend recording during electrocautery.

Respiratory gases and ventilation: Keep anesthetic concentration and $PaCO_2$ constant during phases of measurement.

Technique: Meticulous recording technique is more important in the OR and the ICU than in the EEG laboratory.

Temperature: Maintain stable temperature of stimulated extremities by wrapping them in blankets.

15. Auditory Evoked Potentials (AEPs) and Brainstem Auditory Evoked Responses (BAER or BAEP)

15.1 Auditory Evoked Potentials

AEPs reflect the sequential activation of the VIII nerve (N. acusticus). Thus, all elements of the auditory brainstem pathway can be monitored (BAER). In this modality, a square wave pulse is emitted by an electromagnet through earphones (click stimulus) of 100–200 μsec duration. The response waveforms as they are recorded by the electrode montage (active = earlobe on the same side as the stimulus = ipsilateral; the reference electrode on the vertex = C_z; the grounding on the fronthead = FpZ) results in seven characteristic peaks which more or less can be attributed to specific sites of origin along the primary auditory pathway (Fig. 103).

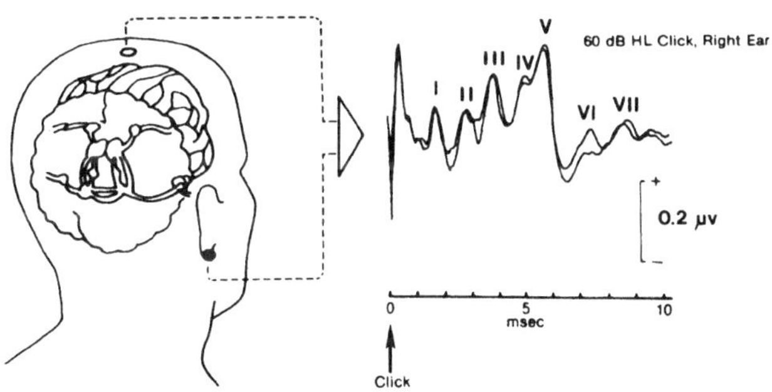

Fig. 103. Typical AEP of the right ear with its seven characteristic peaks reflecting the auditory pathway.

When a stimulus, either a click or a tone burst, is presented to the ear, the nervous system produces a series of responses extending from approximately 1.5 ms to over 500 ms after the stimulus. This sequence of responses can be divided into three general categories on the basis of their latencies and anatomical origins.

The earliest group of responses, occupying the first 10 ms post stimulus, are known as the short-latency auditory evoked responses which arise from the brainstem structures in the rostral medulla, pons and caudal midbrain. The components of the short-latency or brainstem auditory evoked responses (BAER) can be used to measure the functional integrity of brainstem structures, particularly those located in the pons. The BAER can also be used to estimate hearing thresholds in those individuals who are not appropriate for behavioral audiometric evaluations. These may include, but are not limited to, retarded and very young children.

A component of the BAER, wave I of the action potential of the VIII nerve, may also be recorded in greater detail using special ear canal electrodes. This technique is known as electrocochleography and allows detailed study of both the action potential of the VIII nerve and the summating potential which precedes it. Comparison of the relative amplitudes of these two components provides information on the functional integrity of the cochlea.

A second group of responses from the 10–60 ms post-stimulus, are known as the middle-latency auditory responses. These responses are thought to arise from the medial geniculate nucleus of the thalamus and from the primary auditory cortex. While these responses have been of limited use for neurological application, they have been found to be useful for audiological purposes. In addition, recent evidence suggests the middle-latency responses to be of value in diagnosing learning disabilities. A possible derivative of the middle-latency auditory response, the 40 Hertz auditory event-related potential, has recently been shown to have audiologic and neurologic applications. This response is sensitive to lesions involving medial midbrain structures and is also useful for determining hearing thresholds, particularly to low frequency (below 100 Hz) tones.

The third group of responses, extending to 500 ms and beyond, are the long-latency auditory responses. While the precise anatomical origins of these long-latency responses is unclear, they are thought to represent the activity in cortical association areas. Long-latency auditory reponses are used primarily for audiological and psychological purposes.

15.2 Clinical applications of BAER

Brainstem pathology is characterized by abnormal latencies and/or amplitude suppression of peaks generated rostral to the site of dysfunction. Thus, BAER are useful in the following situations (Fig. 104):

a) acoustic pathway lesions caused by tumor or retraction such as during operation,
b) brainstem evaluation such as for brain death,
c) neonatal auditory sensory evaluation,
d) hearing evaluations,
e) demyelinating disease,
f) comatose patients, and
g) operation at the cerebello-pontine angle.

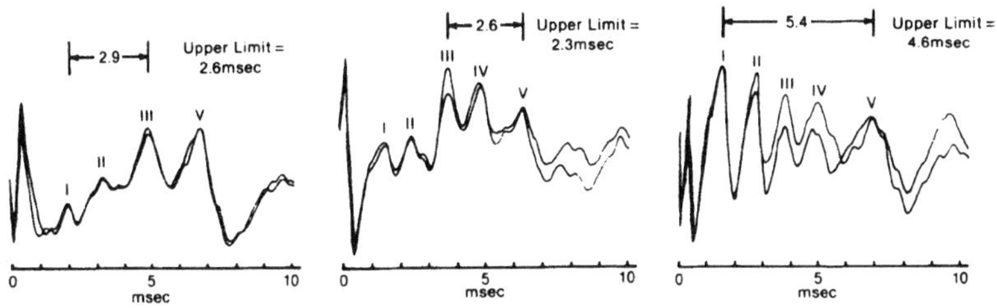

Fig. 104. The use of BAER peak latencies for the diagnosis of sensory dysfunction affecting the VIII nerve. *Left*: acoustic neuroma; *middle*: midbrain tumor; *right*: demyelinating disease. (adapted from Nicolet Diagnostic Series)

As brainstem auditory evoked potentials are subcortical in origin and are little affected by anesthetics, they can virtually be used with any anesthetic technique for monitoring. Absence of BAER in a comatose patient may be due either to neurological dysfunction or to conductive or sensorineural hearing loss. A distinction between the two can only be made by additional recording of the electrocochleogram. If no electrocochleogram can be recorded, one must assume that the absent BAER results from peripheral hearing loss rather than from intracranial pathology.

A correlation between intraoperative changes in BAEP or BAER (brain stem auditory evoked response) and postoperative brain damage fit with the current knowledge about the various generators of the BAEP waveform. Peak I which is generated in the extracranial portion of the auditory nerve and peak II which is

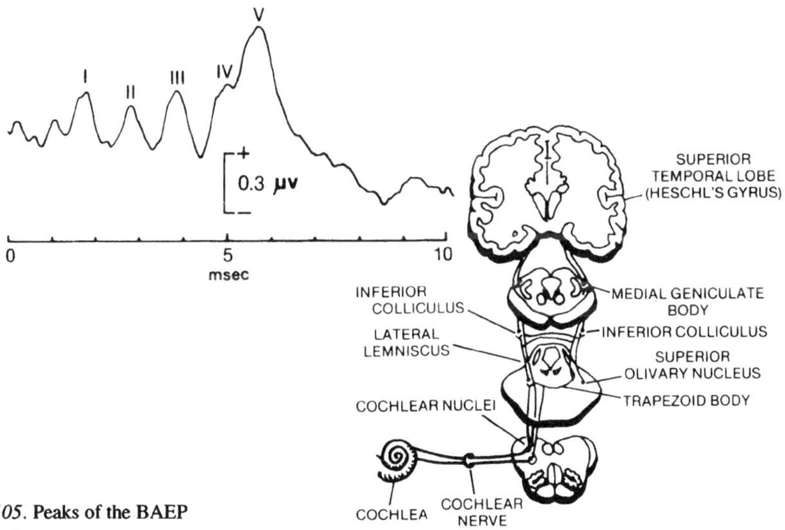

Fig. 105. Peaks of the BAEP and their generators:

I – action potential at the cochlea; II – auditory nerve; III – olivary complex-cochlear nucleus*; IV– brainstem; V – lateral lemniscus; VI – medial geniculate*; VII – auditory radiation*.

* Indicates that the discussion of the origin is still in progress and no final conclusion has yet been reached.

generated in the intracranial portion of the auditory nerve and/or the cochlear nucleus could be preserved even in the presence of an injury rostral to the position in the brainstem (Fig. 105). Thus, one cannot exclude extensive cortical damage even in the absence of alterations in BAEP as they *do not* reflect cortical function.

BAEP are also very insensitive to hypoxia and are normal at oxygen saturations which otherwise will alter the cortical evoked potential [101]. In assessing the value of the BAEP monitoring during operations on the posterior cranial circulation, reports suggest this modality of EP recording to be a sufficiently sensitive device to evaluate stigmata of ponto-medullary dysfunction [102].

The following procedure is advocated in order to get a good signal from the auditory pathway. Using the standard montage (right ear: A_2-C_z, ground FpZ; left ear: A_1-C_z, ground FpZ; Fig. 106) with an instrument setting as listed in the Fig. 106, results should be replicable for each ear and compared among each other for consistency.

If no response to the click is seen on the display,

1. Check whether the stimulus reaches the ear as the earphones have to be over the external ear canal.
2. Check for obstruction or perforation of the ear (can the patient hear the stimulus?).
3. Increase click intensity of the stimulus.
4. Make sure whether the stimulus is transmitted at the cochlea (run an audiogram).
5. Try while switching the polarity of the electrode.

If the noise on the display wave is too intense:

1. Increase the number of averages.
2. Check electrode impedance (set it below 3 kOhm).
3. Twist the electrode cables together as they may act like antennas picking up electrical noise from the environment.
4. Eliminate surrounding electrical noise by turning off nearby electrical equipment, especially all motors, TV monitors, and fluorescent lights.
5. Administer a sedative to relax the patient.
6. Use alternating clicks.

During the measurement the various peaks should read as follows:

1. Wave V is the most prominent and appears around 5.5 ms post stimulus.
2. Wave I should have a latency of less than 1.4 ms. It is of no use in contralateral montage.
3. Wave III should have an equilatency between the foregoing wave I and the post wave V.

The following factors may influence the different waves in BAEP:

1. Latency increase by about 0.03 ms per dB of decrease.
2. Best results are obtained with a click frequency from 2000–4000 Hz as BAEP reflect 1000 Hz and up.
3. A repetition rate of 10 per sec gives the clearest waves. A faster repetition rate can *decrease* amplitude and recognizability. Also, interpeak-latency (IPL) increases slightly at higher rates.

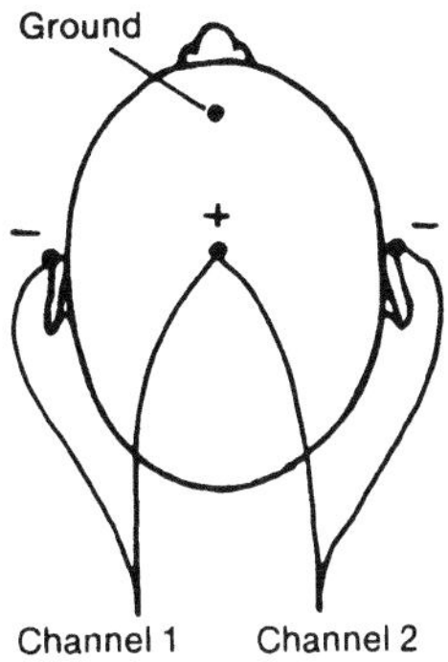

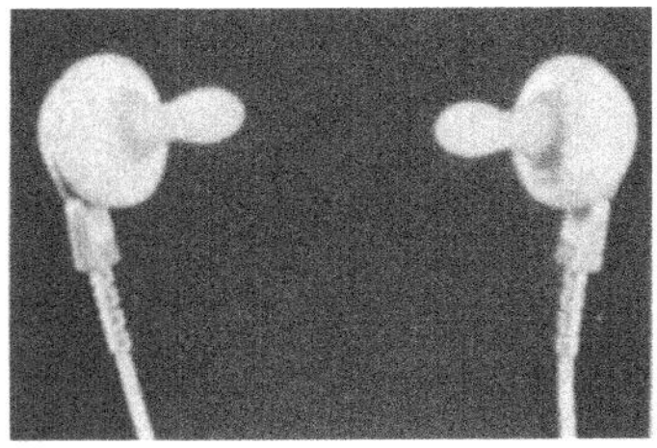

Recording:
Channel 1: C_z–A_1
Channel 2: C_z–A_2
Ground: FpZ

Filters: 30–3000Hz
Sensitivity: ± 25μVolt
Full scale
Sweep time: 10.24 msec
Sampling rate: 50 KHz
Repetition per average: 2000

Stimulation:
Ear-insert transducer
Clicks: Alternating polarity
60 dB SL (personal threshold), 200 μsec duration
Rate: 11.2/sec
Contralateral masking
Noise: 35 dB HL (average threshold)

Fig. 106. Brainstem auditory evoked potentials (BAEP). Placement of recording electrodes and stimulation parameters.

4. Hearing aid stimulators are more useful in the OR and ICU setting than headphones.
   They, however,
   a) intensify sound,
   b) resonate differently than headphones, and
   c) show an absolute delay of around 1 ms due to the tubing.
5. The bone vibration which comes with the click sound by-passes any conductive loss. Because of this bony vibration, a masking noise is needed in the *opposite* ear while stimulation on one ear will be picked up.

Factors which will affect the response include:

a) *Body temperature*: Absolute and interpeak latencies increase as the body temperature drops (0.17 ms/ °C). If the temperature falls below 27 °C, *all* waveforms are lost.
b) A peripheral hearing loss will extend the absolute latencies and decrease amplitudes.
c) Drugs in general have no effect except all *ototoxic* drugs (i.e. gentamycine) which degrade the AEP.

An analysis of the AEP will be devided into:

a) *Audiological*, with absolute and left to right latencies of wave I, while a detailed evoked response audiometry requires measurements of latency changes or the 40 Hz test (see above).

b) *Neurological*, with interpeak latencies (stimulus to peak), left to right interpeak latencies differences, and a computation of the conduction times (CCT) of wave I to V, I to III (peripheral conduction time), and II to V (central conduction), which will be compared with normative values.
c) The *absolute value of wave V*. An additional audiological inquiry determines whether a sensorineural or a conductive loss is present. The interpeak latency of wave V to I should read less than 0.5 ms.
d) *The* P_{35} *component* indicates cortical arrival. It is the major peak of the middle latency response.

16. Visual Evoked Potentials (VEPs)

The purpose of this type of evoked potential measurement is the noninvasive evaluation of the functional state of optic pathways between the retina and the visual cortex. It is useful in the search for organic evidence of a deficit (Fig. 107).

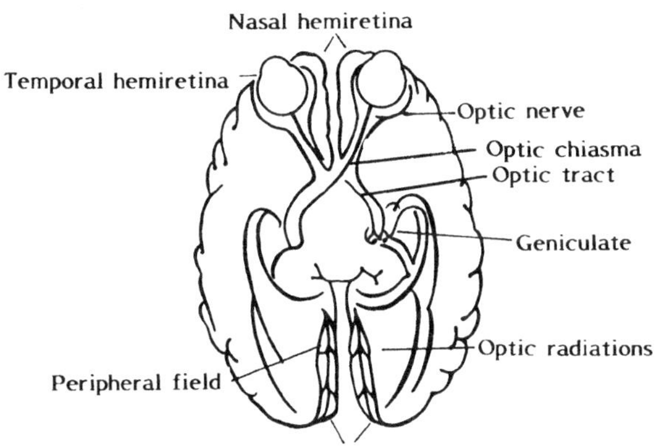

Fig. 107. The physiology of the visual pathway used during VEP monitoring.

Thus, the VEP is helpful during operations which in any way may impede the optic pathway to the occipital cortex such as:

a) hypophysectomy,
b) resection of retro-orbital lesions,
c) procedures involving the occipital cortex, and
d) epilepsy surgery.

For stimulation of the visual sensory path a special flash-stimulus generator in a set of monocular or binocular goggles which are placed over the patient's closed eyelids are used (Fig. 108).

The cortical function dictates that the majority of the visual cortex reflects foveal perception. The response waveform is recorded from the following electrode

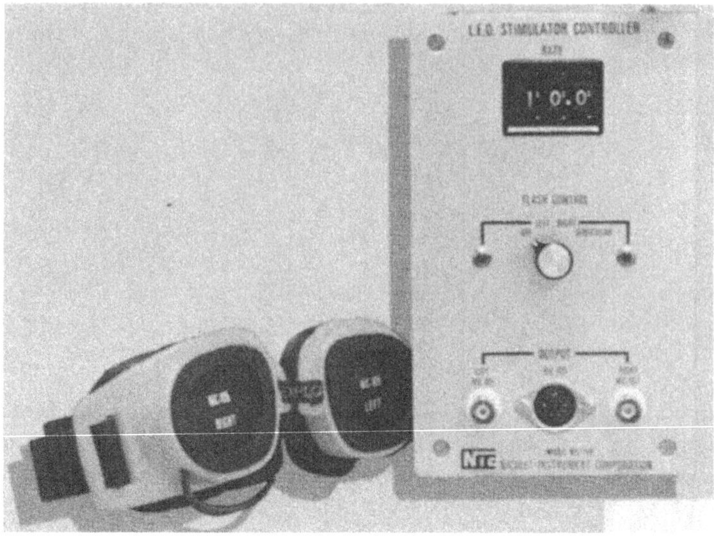

Fig. 108. Stimulus source for the recording of VEPs with variable stimulus rate, using monocular or binocular stimulation.

montage:
- active: O_1 'or O_2'
- reference: C_Z (or A_1 and A_2, respectively)
- ground: FpZ.

With an averaging rate of 100 sweeps, a stimulus rate of 2–11/s and a duration of 5 ms, two detectable major downward deflections named P_{100} and P_{200} are consistently seen.

The preceding N_{70} often is difficult to identify and the later N_{169} is inconsistent in its appearance. All waveforms are generated from the primary visual cortex, before higher association areas. No sub-cortical potentials have been found yet (Fig. 109).

VEP measurements should be undertaken as follows in order to be able to evaluate a potential deterioration in the visual pathway:

1. Determine the difference in latency of the P_{100} obtained from the left and right eye.
2. Determine the difference in amplitude of the P_{200} obtained from the left and right eye.
3. Determine the absolute latency of the waveform components.

Usually, as in the other modalities of the EP, results should be replicated twice and evaluated for consistency. If the waveform, however, tends to be unclear:

a) the ambient light source is to be reduced since any decrease in light contrast results in an increase in latency and a decrease of amplitude;

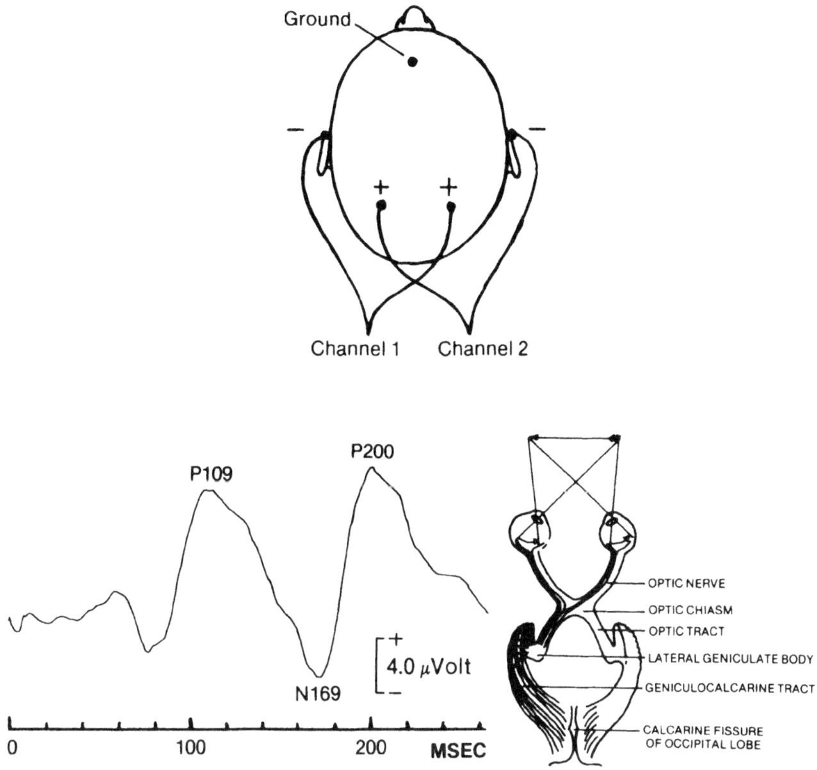

Recording:	Channel 1: O_2–A_1	Sensitivity: ±50 μVolt (Full scale)
	Channel 2: O_1–A_2	Sweep time: 262 msec
	Ground: FpZ	Sampling rate: 3906 Hz
	Filters: 1–100 Hz	Repetitions: 128

Fig. 109. Recording sites used for VEP and the typical waveforms as they are recorded.

b) reduce ambient noise by increasing the number of averages, or increase the relaxation of the patient, or use a 60 Hz notch filter;
c) the replication rate of the foveal stimulation has some impact on waveform configuration. An increase of stimulation from 1–4 Hz results in an increase in latency up to 45 ms. A rate between 8–10 Hz, however, merges to a steady-state response.

Investigators who initially were enthusiastic about the value of intraoperative VEP monitoring, however, are somewhat disillusioned. There are the technical problems that continue to limit usefulness of VEPs. The flash evoked potentials elicited by light-emitting diodes mounted in opaque goggles are *less well* defined and less reproducible than the pattern reversal checkerboard evoked potential (PREP) which is used in the diagnostic laboratory. In the latter the individual has to sit in front of a TV screen focusing on checker board pattern which is reversed every 20–30 ms

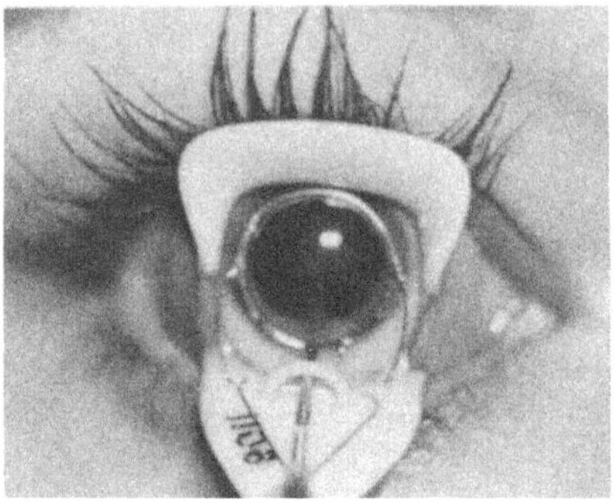

Fig. 110. Electrode placement for clinical electroretinography (ERG).

in regard to black and white contours (pattern reversal stimulus generator). Optic nerve alterations can only be separated from retinal abnormalities by means of the ERG (electroretinogram), an additional visual electrodiagnostic system (Fig 110). Post-chiasmal and chiasmal pathologies may only be detected by half-field VEP, while the full field VEP can localize optic nerve pathology.

The response during VEP is not easy to read as it represents the additive sum of all responses from the visual cortex. Thus, responses from healthy nerve fibers may *mask* responses from unhealthy fibers. As some visual field deficits look like multiple sclerosis, one should not over-read the results when interpreting an increase in latency. One has to consider the morphology as well. Another problem stems from the opaque goggles used for emitting flash signals. The goggle housing should be well supported by the bony ridge of the orbit, as it inadvertantly may slip into a position resulting into direct pressure on the globe of the eye which may cause central retinal artery thrombosis.

Preclinical specifications for intraoperative VEP monitoring including stimulus rate, electrode location, or filter setting have not yet been fully defined. As most of the currently used anesthetic techniques induce small pupils, the reliability of monitoring may be hampered. Additionally, intraoperative VEPs are extremely sensitive to the effects of anesthetic agents [103].

Besides detecting alterations, VEPs correlate well with improvement in vision after operation [104]. Intraoperative irreversible VEP changes were found to go hand-in-hand with postoperative decreased visual function or an intractable intracranial hypertension [105].

Summary on the application of intraoperative evoked potential monitoring

Auditory Evoked Potentials (AEP)

Brainstem Auditory Evoked Potentials (BAEP):
- Acoustic neurinoma resection
- Posterior fossa operations
- Diagnosis of brain stem damage

or

Intermediate and long-latency Auditory Evoked Potentials:
- Resection of temporoparietal lesions
- Epilepsy surgery
- Localization of the specific auditory cortex

Somatosensory Evoked Potentials (SEP)

Short-latency Somatosensory Evoked Potential (SSEP):
- Peripheral nerve lesions
- Spinal cord lesions
- Operative treatment of scoliosis
- Brain stem damage

or

Intermediate and long-latency Somatosensory Cortical Evoked Potentials (SCEP):
- Intracerebral aneurysm clipping
- Hypoxemia, malperfusion
- Effect of central active drugs
- Efficacy of central analgesics
- Prognostic value and efficacy of treatment of head trauma
- Intracerebral bleeding
- Epilepsy surgery
- Stereotactic thalamic procedures

Visual Evoked Potentials (VEP)
- Hypophysectomy
- Resection of retro-orbital lesions
- Operative procedures involving the occipital cortex

17. Complications that arise during EP monitoring

Risk of electrophysiological monitoring of the nervous system is minimal and may be compared to the risk of recording an ECG. Hazards of burn when patients are attached to multiple electrical device are minimized by observing the standard procedures of care when using electrical devices. Any leakage of current from the monitor to the patient should be less than 10 μA, and from the monitor to its case less than 100 μA. It is of importance to make sure that all electrical devices attached to the patient derive their power from a single insulation grounding. One has to keep in mind the possibility of pressure necrosis whenever electrodes are attached percutaneously to comatose or anesthetized patients over a longer period. Especially during VEP monitoring, care has to be taken to avoid pressure from the goggles used to emit flash stimuli.

Last but not least one should be aware of collecting erroneous data due to equipment failure or due to errors in methodology, technical difficulties, or unexperienced personnel. The frequency at which wrong data are collected seems to be inversely related to the experience of the team and, the degree of meticulous attention to technical details of recording and interpretation.

Although the complexity of electrophysiological monitoring in some institutions may reach the boundaries, difficulties encountered in many ways reflect those seen typically during the early days of ECG recording. Some of the systems on the market today which are able to record all modalities of evoked potentials are too bulky, expensive, and not easy to operate. However, the scene is slowly changing as more and more systems get less expensive, incorporate less cumbersome technique, are more compact and flexible, thus opening a new world of monitoring for the clinician.

One shortcoming in EP monitoring is not that of equipment, but that of lack of experienced personnel which is trained in recording and interpreting EPs in the OR and in the intensive care unit. While an EEG technician should be available in the institution for electrode montage, for the meticulous attention of electrode setting as well as data acquisition, the physician is responsible for data interpretation. As the interpretation of EP waveform rests primarily on *pattern recognition,* EP changes have to be looked at in the light of *all pharmacological and pathophysiological* changes.

Regardless of the physician's speciality, the individual involved with the interpretation of EPs would have to gain additional knowledge in an area outside his or her speciality. The skill in EP interpretation in the OR and/or the ICU is in direct proportion to the individual's degree of interest and experience in this new neurophysiological methodology.

18. Systems of use for EP measurements in the OR and in the ICU

Among the various systems on the market which can be used for evoked potential monitoring, few tend to be suitable in adding information with regard to cerebral function. Any physician responsible for the interpretation of EP signals and present in the OR throughout the whole procedure also should have knowledge of the monitoring of other body functions.

Additional cerebral monitoring is only feasible when:

- the basic function for recording is standardized;
- the device is compact and small enough to fit into the room;
- the unit can easily be transported from one room to another;
- the monitor incorporates additional features for cerebral recording;
- the data are represented in an interpretable format;
- the unit incorporates the option of computer-based data storage; and
- the monitoring device and its function are backed up sufficiently by a service department of the company.

Some of the more sophisticated systems such as the Nicolet Pathfinder™, the Amplaid MK 15™, the Cadwell 8400™, or the Neuropack 8™, for instance basically are instruments meant for detailed neurophysiological and/or purely neurological diagnosis finding. Among those devices which may fit into the scope of the physician who is interested in the evaluation of central nervous function intra- and postoperatively, smaller systems with lesser options and capacity for data evaluation and with lesser costs are sufficient to answer all those questions which may arise in an operative environment. Among the systems which nowadays are available on the market, and which more or less incorporate a number of features to help in the evaluation of function of the CNS being at stake, are as follows:

1. The Neurotrac™ (Interspec Medical, Wisconsin, USA).
2. The Lifescan™ (Neurometrics, San Diego, Calif. USA).
3. The Traveler™ (Biologic Systems Corp., Illinois, USA).
4. The Nomad™ (Tracor Northern, Wisconsin, USA).
5. The Cadwell 5200A (Cadwell Lab., Washington, USA).
6. The Neuropack II Plus™ (Nihon-Kohden, Japan).
7. The Amplaid MK10™ (Amplaid, Italy).
8. The Nicolet Compact Four™ (Nicolet Biomed. Inst., Wisconsin, USA).
9. The Evoport (Tönnies/Jaeger Medical Electronics, Würzburg, FRG).

Some of these systems have already been mentioned in the previous chapter on EEG monitoring (pp. 68–83). Thus, they provide the user with EEG data analysis plus the additional option of evoked potential monitoring. Among the systems incorporating EEG as well as EP monitoring, three of them distinguish themselves because of their compactness and ease of handling which gives them a good chance of being used regularly by the clinician not solely involved in neurophysiological measurements. As many measurements are done in the OR and/or the ICU, the speed and the ease of transportation is another point which is necessary to consider when the decision comes up to buy such a monitoring device.

The Neurotrac™

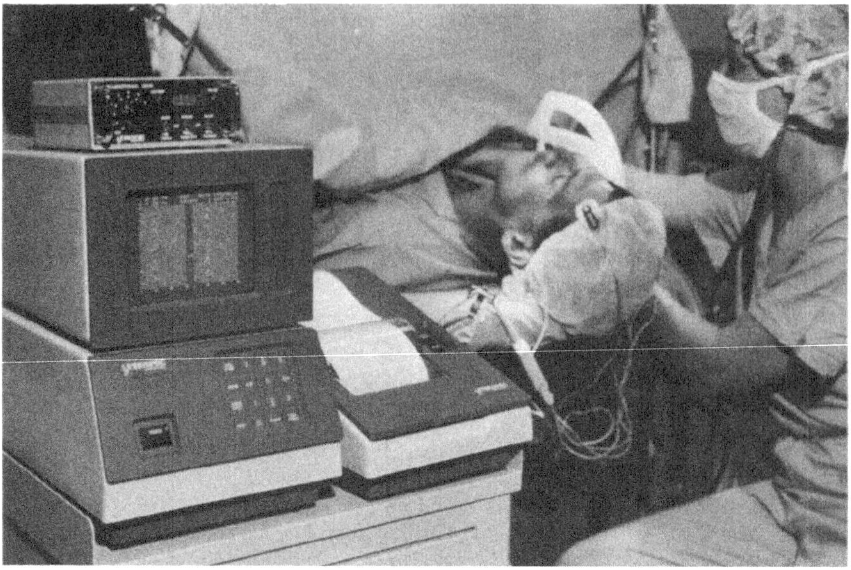

Fig. 111. The Neurotrac™ in use in the OR.

Features of the evoked potential include:

1. Two channel averaging.
2. Six trace storages and recall.
3. Menu selection of averaging parameters.
4. Superpositioning of up to six traces possible.
5. Two independently movable cursors.
6. Traces may be printed in graphical format with all pertinent data.
7. Waveform peaks marked graphically and values printed in tabular form.

Features of EEG processing include:

1. Compressed Spectral Array display.
2. Spectral Edge Frequency, the outer marker of the CSA display for the recognition of changes in the higher frequency levels in relation to power.
3. Spectral (bar-graph) histogram display format in half-Hertz resolution for a more detailed information.
4. Power band display format for quantification in four selectable EEG ranges.
5. Dot clustering with total power of the EEG.
6. Storage capacity of 10 full screen displays for recall at any time for comparison with current data.
7. RS-232 interface to permit export of raw and spectral EEG data and evoked potential waveforms for long term storage or additional analysis in a host computer.
8. Automatic print-out of data via a separate printer.

The Lifescan™

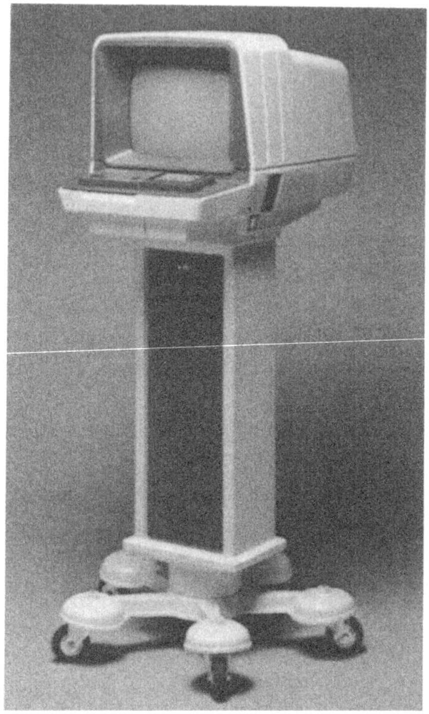

Fig. 112. The Lifescan™ as it comes on wheels for mobility.

This system comes with wheels permitting fast move from one place to another. In the EP mode this system enables the user to:

1. Analyze EPs concurrently with EEG aperiodic analysis.
2. Analyze SEP curves in color creating a distinct and easy form to differentiate images even at some distance.
3. Select from up to four channel averaging with display of information from two channels.
4. Storage trace memory of six selectable previous traces.
5. Superimpose traces in different colors thus permitting easier identification of recalled EP data.
6. Use two independently movable cursors for peaks identification and calculation.
7. Print traces and quantitate information of recallable traces from the color display screen by a built in printer.
8. Mark waveform peaks in multiple colors for faster and easier identification of peaks and valleys of EP waveform.
9. Use a remote control infrared communication for operation of display options.

In the Aperiodic Analysis mode the following features can be used:

1. Real-time wave-by-wave analysis of the EEG.
2. Delineation of power in the various frequency bands by the use of different colors.
3. EEG power display in a three-dimensional fashion.
4. Expanded memory permitting recording of 10 hours in the 10 minute and 24 hours in the 30 minute mode.
5. Six user-selectable snapshots of EEG power data storable for later comparison.
6. 20 minute memory protection for transportation with the patient while recording goes on.
7. RS-232 interface permits communication with a separate laptop computer (Kaypro^{R} 200 PC) for research work with a separate system-adapted software, doing quantification of EEG data, displaying histograms and storage of data on floppy disk for later retrieval.
8. Hard copy capabilities by the built-in printer.
9. Spectral Edge Activity for easy detection of changes in the fast frequency domain.

The Biologic Traveler™

Fig. 113. The portable Traveler™ due to its capabilities permits extensive neurologic monitoring even in remote places.

In addition to EEG frequency analysis (using FFT) the multi-modality four-channel system can be used for

a) auditory,
b) visual, and
c) somatosensory evoked potential measurements.

For a more detailed neurologic examination

d) ENG (electroneurography)
e) EOG (electrooculography)
f) ERG (electroretinography), and
g) 40 Hz ERP (event related potential), as well as
h) ECochG (Electrocochleography) can be implemented.

The system can be expanded to include topographic brain mapping (see later). In the evoked potential mode the following features are incorporated:

1. Parameters can be defined via a keyboard with which the menu is driven.
2. Dual floppy disks for permanent data storage and storage of patient records, test parameters and comment.
3. There is a 100% IBM PC compatibility.
4. A separate black and white thermal pritner permitting plotting of data and evoked potential waveform.
5. Up to eight superimposable waveforms.
6. Two interactive cursors with readout of latency, interpeak latency, and amplitude (peak to peak and base to peak).
7. Up to 300 waveforms stored on the floppy disk.
8. Storage capacity increase up to 10 Mbyte by an optional hard disk.

The EEG data analysis incorporates all of the features (spectral edge frequency, power bands, histograms, CSA, etc.) as most of the other devices.

The Nomad™ Diagnostic System

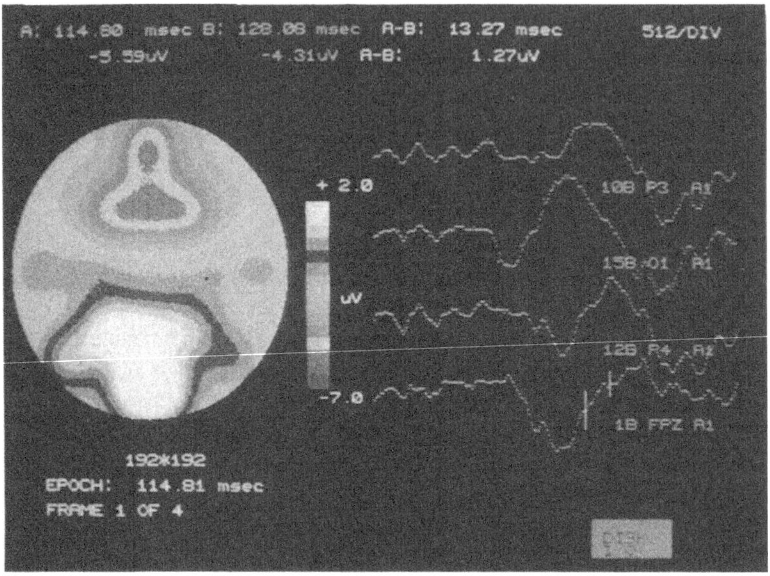

Fig. 114. The Nomad™ system with its expanded capability of field summation Topographic Map Display.

This device offers a wide range of data acquisition, processing, and storage capacity. The system, however, is more rugged and demands more space during practical use. Due to the wider range of options, when compared with other instruments, the operator needs some time to get accustomed to.

The systems capabilities are as follows:

1. a) auditory,
   b) visual, and
   c) somatosensory evoked potential monitoring.
2. Floppy and hard disk storage for data, protocol parameters and subjects reports.
3. A 16-bit processor giving the system more speed and multitasking capabilities.
4. A 10-bit A/D conversion offering four times the amplitude resolution of a competitive 8-bit system.
5. Simultaneous acquisition of four channels of data.
6. Multiple, *independent* somatosensory stimulation.
7. Data processing including addition/subtraction and waveform inversion.
8. A large 12-inch high resolution color graphics monitor for display of up to eight waveforms including cursor readout recording parameters and waveform latencies.
9. A separate ink-jet printer for comprehensive subject reports, subject information, waveform data, and recording parameters
10. Aside from the evoked potential mode, capabilities of implementing into the system a field summation Topographic Map Display (Fig. 114), which sums up the electrical fields from 16 electrodes.
11. The additional feature of CSA during EEG processing may be expanded to CDSA (continuous density spectral array) to prevent low level of EEG activity in one power band to be obscured by high-level activity of the same frequency in the conventional CSA. Thus, each epoch of data is displayed in a line of varying color comprising three parameters, time, frequency, and power.
12. Quantification of EEG data includes most of the common modalities such as:
    a) spectral edge frequency,
    b) dominant peak tracing,
    c) frequency band ratios,
    d) per cent power of the total spectrum, and
    e) magnitude of each frequency band.

The Cadwell 5200 A

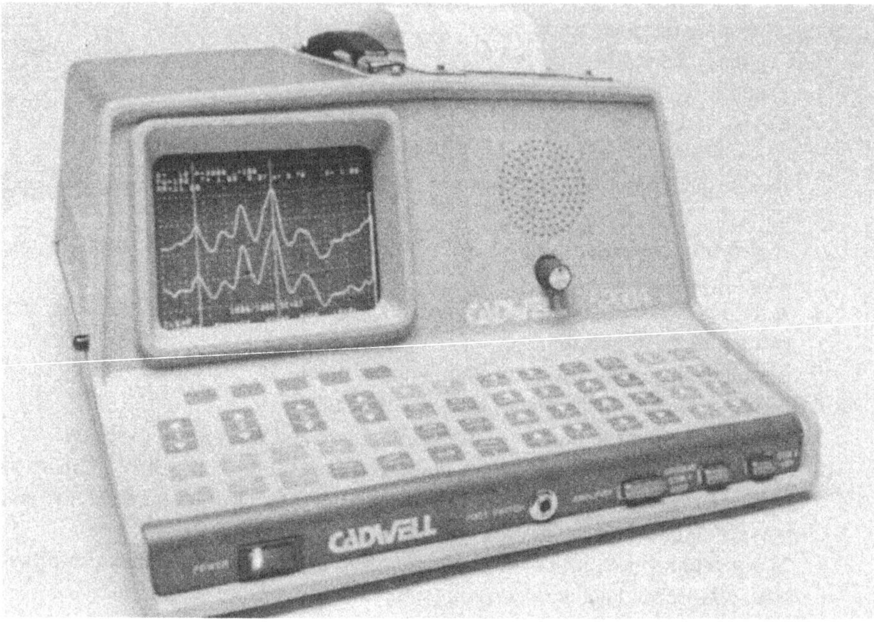

Fig. 115. Front view of the Cadwell 5200 A for expanded neurological testing.

Another two-channel evoked potential portable and compact analyzer with four storage capacities and a built-in X-Y-plotter to record the data, as they are displayed on screen. The following capabilities are incorporated:

1. Usage for SEP, VEP, and BAEP evaluation.
2. Continuous spectral array presentation.
3. Additional options for neurologic assessment with a full set of function such as:
   a) EMG (electromyography),
   b) NCVS (nervous conduction velocity of sensory nerves),
   c) NCVM (nervous conduction velocity of motor nerves), and
   d) single fiber analysis.
4. Remote operation with a RS-232 interface.
5. Special options include:
   a) a 5 1/4" hard disk storage system (10 Mbyte) for storage and faster retrieval of data,
   b) topographical mapping of electrical activity of the brain in 16 colors (for details see last chapter on brain mapping) permitting motion picture effects of EPs as it travels in time,
   c) topographical mapping of Compressed Spectral Analysis of the EEG in color in four arrays according to the frequency bands selected, and
   d) a 16 color printer for brain map printouts with color scaling and labelling of maps.
6. For the evaluation in audiology/neuro-otology
   - the 40 Hz Evoked Response Audiometry
   - the ENG (electronystagmography), and

   for further assessment of the visual pathway
   - ERG (electroretinography, scotooptic and photooptic)
   - Flicker fusion frequency, and
   - the EOG (electrooculography)

   can be implemented.

The Neuropack-II Plus™

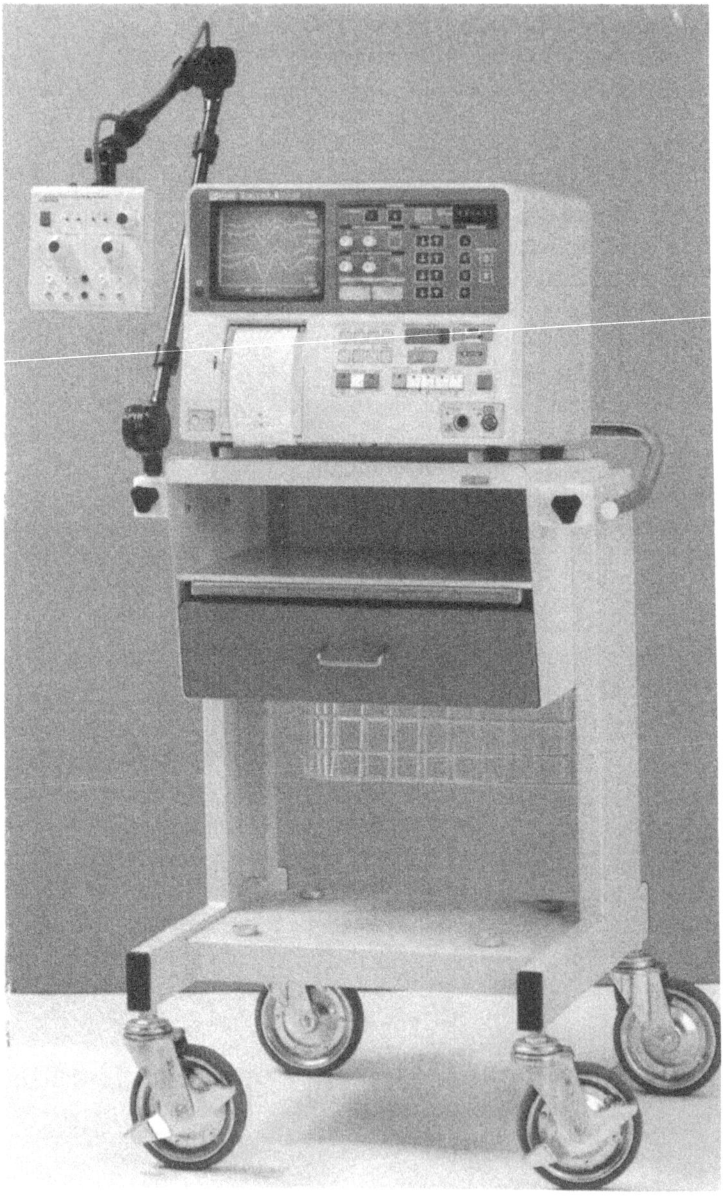

Fig. 116a. Front view of the Neuropack-II Plus™, which can be used for all modalities of evoked potential testing.

Among those evoked response recorders which can be used for all modalities this device is characterized by its comprehensiveness and ease of use.

It contains a total of 16 programs of elective somatosensory, auditory, and visual evoked responses, each with three different preset programs and manual programming.

Placed on a movable cart the system contains the provision for a monitoring plotter and an optional unit for an EMG loudspeaker and LED stimulator.

Implemented are two channels for EMG as well as sensory and motor nerve conduction velocity measurement. EEG power spectra and mapping cannot be assessed with this type of monitoring.

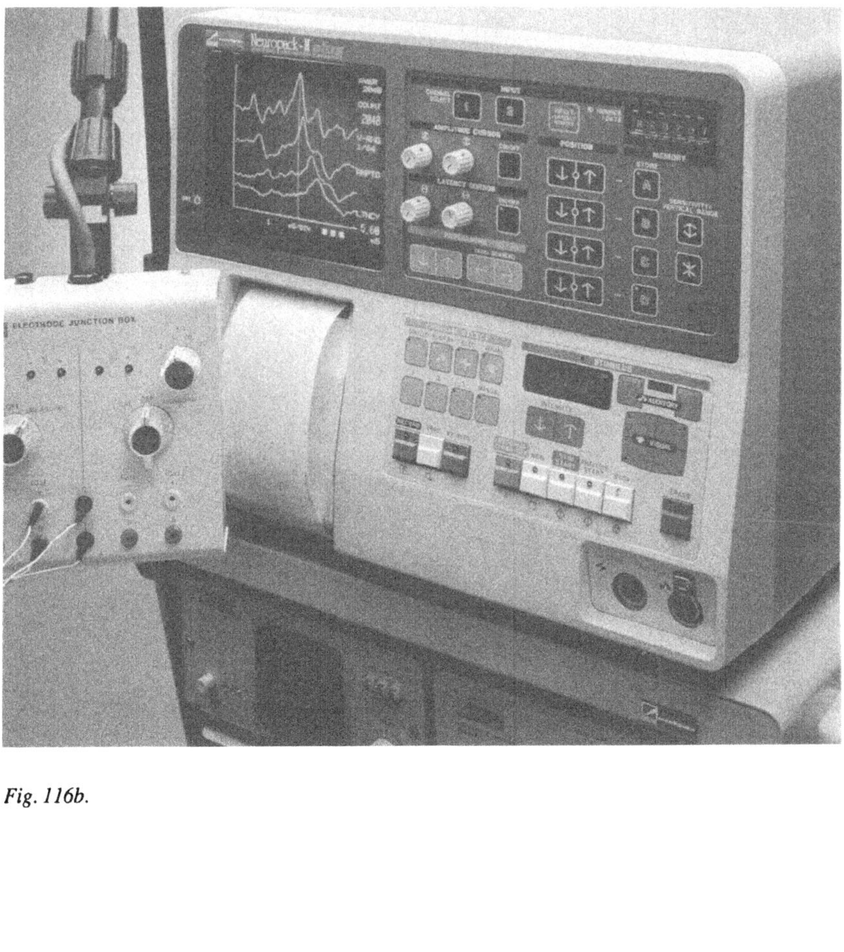

Fig. 116b.

The Amplaid MK 10™ and MK 15™

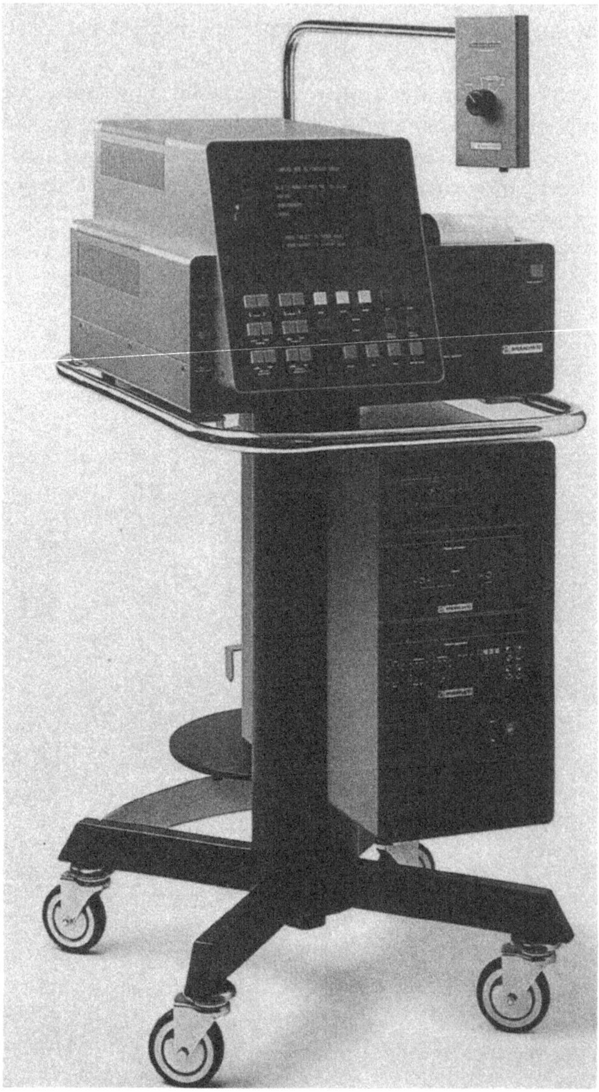

Fig. 117a. The Amplaid MK 10™ as it is mounted on a trolley with a separate storage system.

This monitoring device is a multisensory potential unit and similar to the Neuropack, can only be used for evoked potential recording. Evoked potential monitoring includes:

a) auditory,
b) somatosensory,
c) visual sensory, and
d) neural pathway testing.

The unit automatically sets up suitable stimulation and acquisition parameters for any selected test. Operation, similar to the previously mentioned device, is a simple matter thanks to the clearly defined functions of the keyboard button. Although the nine acquisition parameters have been preset according to the most commonly used values, they can be modified by the user. A two memory screen is available for visualization and validation purposes. It is provided with a trolley thus permitting flexibility in location changes. One interesting feature of the signal processor system is its comparably low cost.

Fig. 117b.

The Nicolet Compact Four™

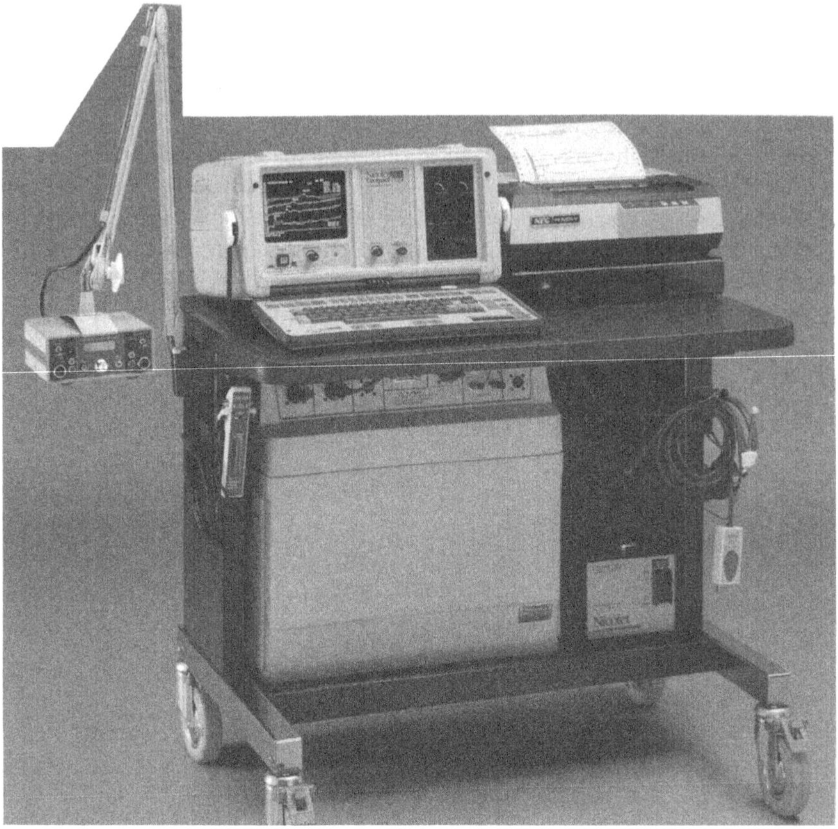

Fig. 118a. The Nicolet Compact Four™, a multidiagnostic evoked potential system with printer and central processing unit.

It is the smaller brother of the worldwide-known Nicolet Pathfinder™, a neurological function test system implemented in many EEG departments. The system is specially designed for

a) somatosensory,
b) auditory, and
c) visual pathway evaluation.

A separate printer comes with a trolley for convenient change of locations. EMG testing can also be performed involving triggering of motor unit activity.

The full 128 character keyboard is used for the entry of complete patient information and comments about waveform and data. The keyboard is also used to define the test parameters, electrode montage, while impedance testing as well as the control of all system functions is directed through specially designated buttons. Operated as a software-based system, all programs reside on a floppy disk which can also be placed in the program disk drive. Due to the software-based capabilities, enhancement and new innovation in electrodiagnostics can easily be implemented by installing a new diskette. Approximately 200 evoked potential waveforms can be stored on a single data disk, handled through the use of a single filing system which is based on the actual patient name, including stimulus parameters, recording parameters and comments.

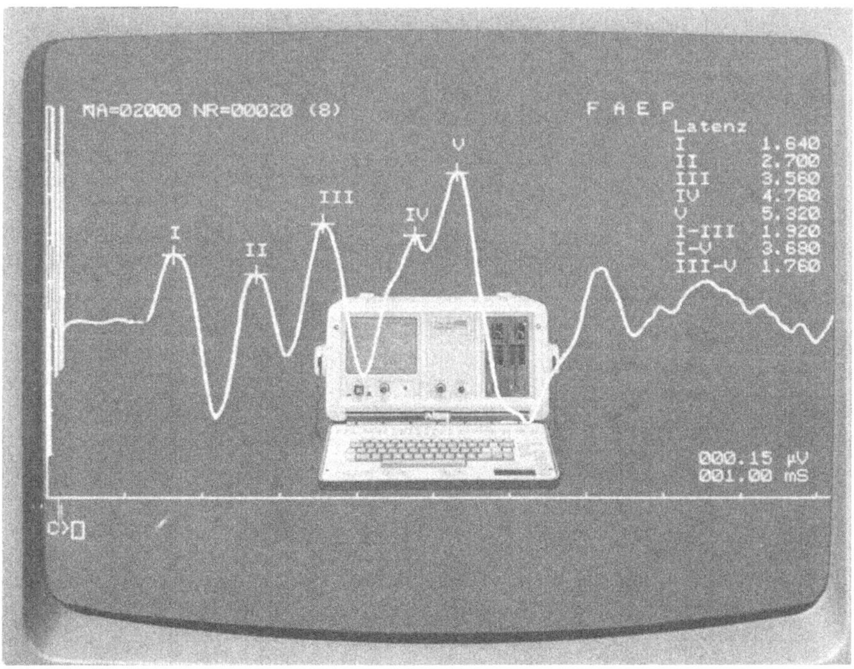

Fig. 118b.

The Evoport 200™

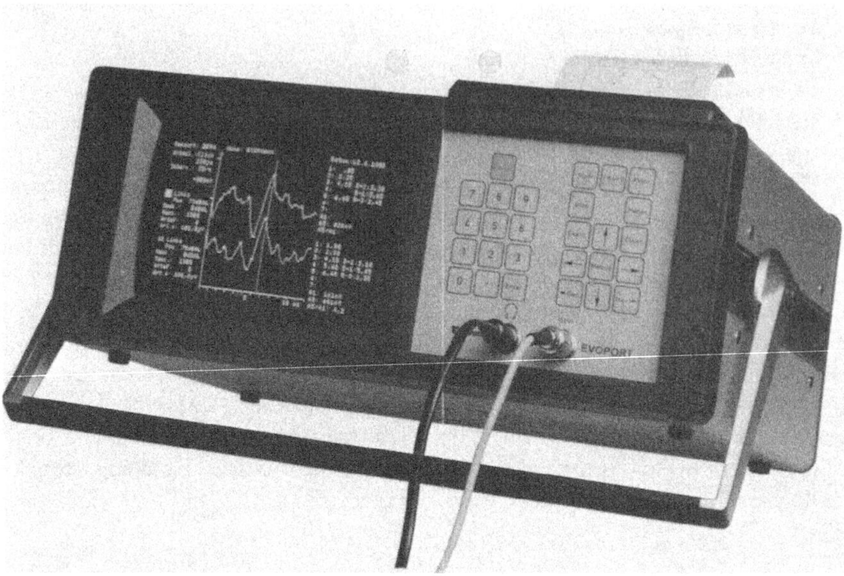

Fig. 119.

A real lightweight and portable unit which offers 'only' flash VEP and AEP analysis. The latter is subdivided in:

1. BERA (brain stem evoked response audiometry) for the measurement of potentials of the auditory nerve on its way through the brain stem up to the acustic brain center.
2. MERA (middle latency electric response audiometry) measures the potentials of medium latency found within the 10–50 ms window. Their general application, however, is limited as the potentials depend, on the respective muscle innervation leading to vast deviation.
3. CERA (cortical electric response audiometry) for the examination of late acoustic response, the so called cortical response. They pick up an area lying on the highest point of the auditory pathway, directly on the cortex.

BERA as well as CERA are very insensitive to anesthesia and sedatives, and are of great aid in evaluating hypoxic changes involving the brain-stem.

19. New scopes in cerebral monitoring by topographic mapping of EEG-power spectra and EP waves

In recent years a development has taken place in the field of electroencephalography, which has expanded the possibility in looking into time-related changes of EEG and EP waves.

By breaking down the EEG into its frequency components and displaying the results in the form of color maps, topographic mapping has succeeded in making EEG interpretation over the whole cortex more rapid and quantitative. The technique was only feasible with the latest computer technology combined with a sophisticated software and a number of special features which enable the user to get a complete picture of the whole cortical surface.

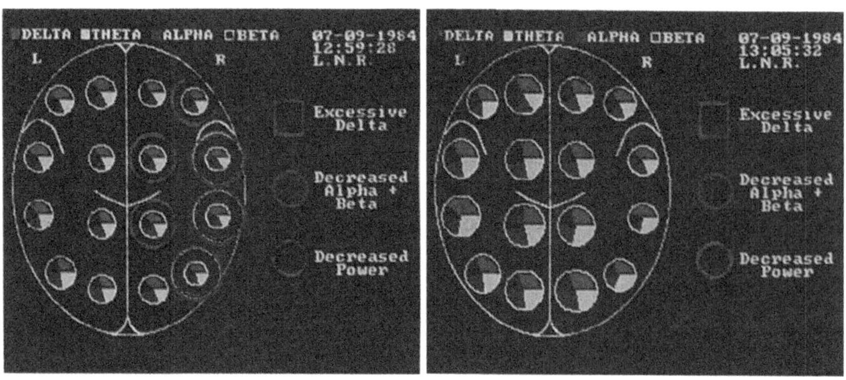

Fig. 120. Example of EEG power mapping, where the different frequency bands are characterized by different colors over the cortex. Warning limits are set by a 50% reduction in the alpha/beta domain indicating hypoxemia.

Left: 3 Minutes following occlusion of the right carotid artery, the power in the six right hemisphere channels has decreased below the warning parameter settings. The Cerebral Tracer™ automatically displays a power decrease warning for the six channels, and an alpha/beta frequency decrease for 3 channels.

Right: 8 Minutes following shunt placement, warning parameters no longer appear on the circles. The power has increased above the warning level compared to the patient's baseline. Right-left power symmetry has improved substantially.

The distribution of various frequencies is indicated by various colors while maps are updated as rapidly as every half second during real-time analysis. Maps can be viewed from different projections such as coronal, left or right sagittal and left minus right sagittal. The multiple maps display capabilities allowing the user to view maps of five frequency bands simultaneously with the upper and lower limits of the frequency bands specified separately (Fig. 120).

Thus, the clinician is able to detect asymmetrical activity or even the focus of abnormal activity. A high resolution color ink-jet printer allows the summary of the patient's recording to be stored for permanent record keeping. Additionally, the computer is able to compare EEG changes within the patient himself taking selectable baseline periods as control. Such an expanded mapping of EEG power activity is meant not only for research purposes, but also for the continuous on-line monitoring in the OR to detect intracerebral ischemia which automatically will alert the user of the patient's EEG changes. The significant advance over a two-channel computerized monitoring is that with 8, 16, or even 32 channel recording one

- has a more complete brain coverage,
- achieves recognition of focal changes,
- differentiates ischemic focal changes from global anesthetic changes,
- has a greater discrimination of artifactual changes from actual EEG changes with no loss of larger areas of the cortex, and
- has more warning parameters, such as a total power suppression by 50%, an alpha/beta frequency decrease below 50%, and an excessive increase of power in the delta band by 200% above control, all of which are useful settings in the evaluation of malperfusion (Fig. 120).

In the above example during right carotid endarterectomy of a 79 year old male, hypoperfusion of the right hemisphere is indicated by the relative drop of power (square over amplitude) of the channel's signal in circle size while the change in frequency distribution is indicated by a relative change of color within the circle.

The following devices are able to perform field summation topography of EEG activity over the whole cortical area and are worthwhile to mention:

1. The Neuromapper™ 1620 or 2050 (Neuroscience, Essex, UK)
2. The Brain Atlas™ (Biologic, Illinois, USA)
3. The Cerebral Tracer™ (CNS Inc., Minnesota, USA)
4. The Cadwell™ 8400 (Washington, USA)
5. The Tracor™ (Tracor Northern, Wisconsin, USA)
6. The Nicolet Pathfinder™ with a mapping accessory for SEP and BEAM (brain electrical activity mapping) for EEG topography (Nicolet Biomed. Instr., Wisconsin, USA).

In addition to EEG mapping, the distribution of evoked potential voltage over space and time on a topographic graphical color mapping of the human brain is also possible. This function provides information, similar to the spatial distribution of the EEG frequency spectrum, of the extent of abnormal activity over time. The detection of abnormalities from a normal distribution pattern based on probability difference mapping of evoked potentials, normal distribution pattern is feasible. For any given test the microprocessor subtracts the patient's results from the normative values, frame by frame, and plots in color the significant voltage differences between the two (Fig. 121).

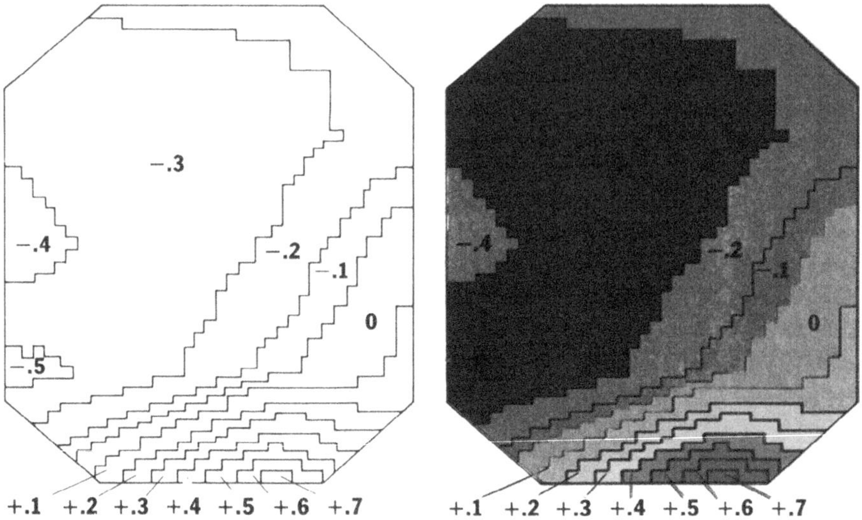

Fig. 121. Example of evoked potential mapping as an enhanced spatial resolution waveform analysis.
Left: The electrical activity over the entire scalp is reconstructed as a contour map by interpolation of voltage values between electrodes.
Right: Each voltage is assigned a color. 'Hot' (red) colors for upward activity, 'cold' (blue) colors for downward activity, which results in a smooth color-contoured map.

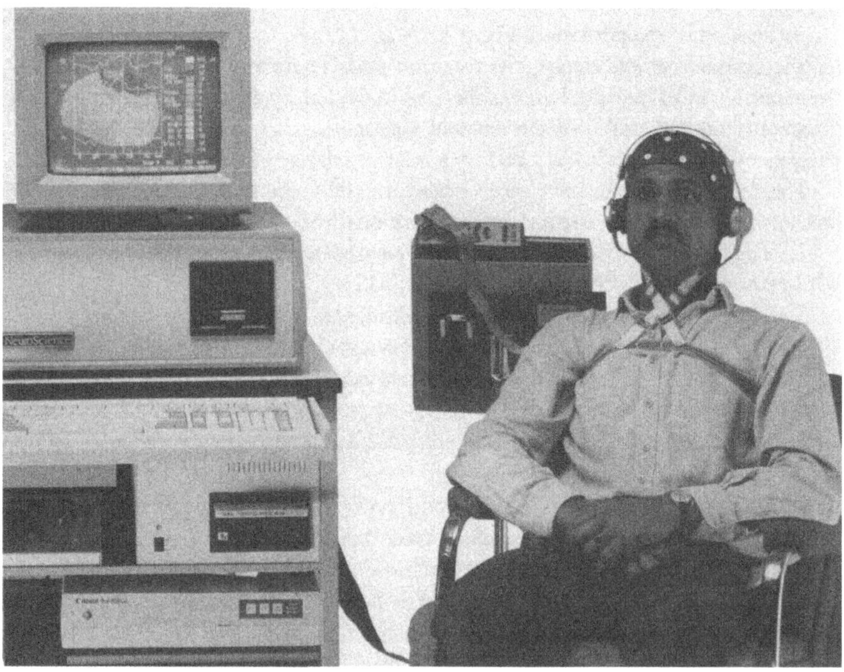

Fig. 122. Set-up for mapping of evoked potential analysis using a swimmer's cap to incorporate up to 32 leads.

The distribution of electrical activity across the whole scalp, however, requires multiple-channel recording to precisely define the site of specific neural generators. Localization of such generators furthers the understanding of sensory information processing and brain functioning. It remains to be seen, whether such a sophisticated extension in computerized EEG and SEP monitoring, which results in the visualization of EEG and SEP maps, will indeed enlarge the scope in CNS monitoring. Some of the potential drawbacks with such an extended device are the multiple (up to 32) EEG and SEP leads which are fixed to the scalp by means of a swimmer's cap, and the size of the instrumentation (Fig. 122). Especially the latter may result in problems when monitoring becomes an essential part in the OR where other electrical instruments are already crowded around the operating table. Additionally, one has to consider costs versus benefits, making sure that CNS monitoring in the OR or the ICU does not become more expensive than a diagnostic laboratory.

20. Appendix

20.1 Care of electrodes

The silver/silver chloride stick-on electrodes are manufactured of pure silver with their surfaces chlorided. They are of circular shape, have a domed top with a small hole in the center, suitable for the application of electrode jelly. A highly flexible PVC-covered cable is attached to the electrode by a crimp connection. The joint is covered by shrink sleeving. Many EEG users could get better results if only a little more attention was given to the care of such electrodes. First and foremost, one must appreciate that electrodes are reusable but have a limited useful life. Nevertheless, their life can be prolonged and their performance considerably improved if the users allow themselves time to prepare used electrodes.

Although claims are made for various metals, or plated metals, giving superior EEG and SEP recording performances, up to this date it is the general opinion of the leaders in EEG recording that silver/silver chloride electrodes, when carefully looked after, are difficult to improve upon.

The EEG and EP electrodes commonly used in Europe and the US are made from pure silver and coated with silver chloride. Such an electrode acts as a reversible chloride electrode and with care can be rechlorided quite easily. The following incorporates some simple hints on the preparation of silver/silver chloride electrodes (Fig. 123). Preparation is much more complex as outlined below; for further information textbooks or publications on practical physical chemistry or electro-chemistry should be consulted (Cooper [106]).

1. Use only pure silver.
2. Clean silver before chloriding with an abrasive such as sand paper, or abresive paste, e.g., jeweller's rouge, but not steel wool or a file, because this can lead to a contamination of the silver with iron.
3. Chloride at a rate of 2.5 mA/sq.cm for several minutes in bromide free sodium chloride solution. The strength of the solution is not critical, but at least physiological saline, and up to 5% concentration should be used. Contrary to general belief, the properties of the chloride layer are not changed by exposure to light. After preparation, the electrodes will show potential differences which may amount to tens of millivolts or even more in some cases.

4. If the electrodes are stored in saline or externally shortened together these potentials can be discharged, but when disconnected or in use, some of the potentials return over a period of minutes or hours. Thus, slow drift is of little consequence if AC-coupled amplifiers are used (i.e. the normal EEG amplifier), but it is of great inconvenience if DC recording is required. Fig. 123 shows a suitable method to silver chloride a used set of stick-on electrodes. They should be cleaned of old chloride with a suitable non-metallic abrasive paste, or a similar scouring powder, washed clean under cold water, and then ideally electrolytically cleaned by 'reverse silver chloriding' at about 20–30 Volts DC, i.e. the recording electrode is made negative by about 20 Volts DC with respect to an indifferent electrode for a few seconds. All stick-on electrodes then are connected to the positive terminal of a 1.5–2 Volt battery. The indifferent electrode may be a piece of silver, or carbon, and is connected to the negative terminal.
5. The electrolyte solution is 5% NaCl and, once used for this purpose, must be discarded. Each electrode is lowered into the saline solution for approximately one minute. It will be seen that the original bright, silver-colored electrode turns into a dark violet color when fully silver-chlorided.
   EEG electrodes perform best when they are equipotential with respect to each other, and remain so even though currents are passing into them and out again. This equipotential is only possible when they are silver-chlorided.
   Re-chloriding is necessary after 30–60 recordings. In practice, even when the electrode looks clean as though all silver chloride was rubbed off, it will still perform well. It is difficult to get electrodes looking evenly coated with silver chloride after they have been used, and although they look mottled and patchy, nevertheless they will record quite satisfactorily.

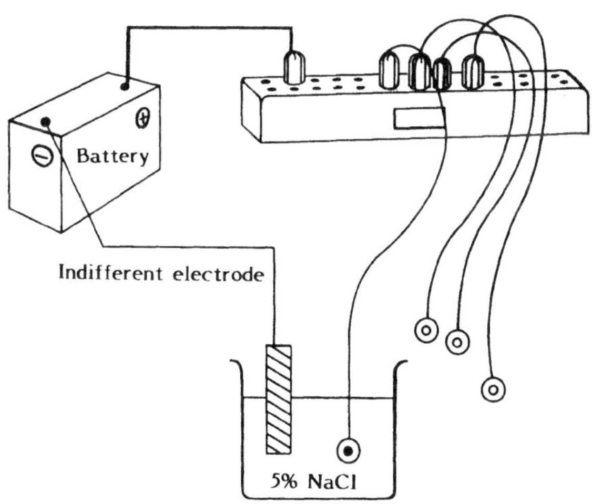

Fig. 123. Methodology for resilver-chloriding of stick-on electrodes.

6. Needle electrodes are made of hard metals. Platinum for sub-dermals, stainless steel for sphenoidals. These metals polarize resulting in different resistances.
   The main reason why the needle electrodes did not become number one in EEG and EP recording is that they have to be autoclaved in order to avoid infection. Normal plastic insulation does not withstand to continuous autoclaving, so these electrodes are considered as unreliable and expensive by most laboratories.

20.2 Summary of the clinical application of EP monitoring in the OR

The range of applications for sensory evoked potentials is growing, as demonstrated in several reviews [103, 104]. Evoked potential monitoring indicates when something is altering the neural pathways, provided that the effect of anesthetics, temperature, and arterial pressure can be excluded as causes.

Visual evoked potential monitoring, which may present a useful technique in operations involving the optic nerve and the chiasma, however, has not had the expected success in application. This is mainly due to the difficulty in delivering a stimulus during surgery. Firstly, unshielded flashing of a stroboscope is distracting to the surgeon, and light-emitting diodes mounted in goggles are not a universal effective stimulus. Secondly, there is the problem of inherent variability of potentials, and thirdly, there is the depressant effect of anesthetics. The variability of the potentials and the inaccuracy of the results do *not* justify routine VEP monitoring. Monitoring of auditory brain-stem evoked potentials (ABEP) has a much greater success, especially in posterior fossa procedures [105].

Most common applications of somatosensory evoked potential testing (SEP) are operations for scoliosis and related disorders, as well as neurosurgical and vascular procedures.

The area of controversy is the choice between spinal and cortical recording. While spinal recording involves the disadvantage of needing epidural electrodes for high-quality recording and the problem of *multiple pathways* being stimulated, cortical recording has the advantage of being non-invasive. The EPs of cortical recording are, however, greatly influenced by hypotension and anesthesia. The recently described technique of transcranial stimulation of *motor pathways* and the recording of large descending spinal pathways via oesophageal electrodes present an extension of this methodology [107]. Clear-cut indications where SEP monitoring is used beneficially are:

- traumatized lesions of the brachial plexus;
- aneurysms of the descending aorta with extensive replants by grafts. This incorporates occlusion of the aorta and hence interruption of the anterior spinal cord supply via intercostal vessels. Time limits for loss of sensory evoked potentials can thus be defined. Additionally, those arteries necessary for nervous function, after reimplantation result in a recovery of the depressed signal, can thus be defined. However, whether the SEPs are representative for the more important motor pathways remains to be seen;

- during carotid artery surgery, EEG monitoring in this respect has some limitations in that an infarction of the middle cerebral artery may result in ischemia of subcortical layers. Thus, warning signs are better derived from evoked potentials which have to travel through the area of ischemia. The EEG will only detect ischemia of the cortical layers.
- *continuous EEG- with interlapped SEP monitoring* which nowadays seems sensible to use as a combined approach. If the signal of SEP averaging plus EEG recording and processing deteriorates against a baseline at the moment of clamping, it would mean the indication for by-pass shunting.
- in cardiac surgery where marked hypothermia presents a problem as cooling may mask ischemia. In view of the lower vulnerability of the brain-stem than the cortex to ischemia, *BAEP* are *inappropriate* for an early warning signal. The *short latency SEP* remains the *most valuable* in this respect. Peaks tend to be attenuated if temperature drops below 25–20 °C
- during controlled hypotension as it is used in the case of operating upon an intracranial aneurysm. SEPs serve as an early warning signal of ischemia in which central conduction time (CCT) shows a close correlation to pressure, especially when the blood flow drops below 30 ml/100 g/min [108].

20.3 Summary of the clinical application of EP monitoring in the ICU

Evoked potential abnormalities have to be interpreted in the light of whether or not they represent a transient dysfunction or a permanent damage. This is the main reason for serial recording of SEPs in the ICU in order to observe trends.

Somatosensory evoked potentials are especially valuable in determining the mechanisms of coma. When serial recording shows a lack of cortical components but a preservation of waves generated in the spine and the brain-stem, cortical lesions have to be suspected.

When cortical potentials are visible, central conduction time (CCT) presents a good prognostic indicator. It is markedly prolonged in patients after head injury, less in non-traumatic patients. A gradual return of CCT to normal ranges in patients with traumatic coma reflects the rate and the degree of improvement being independent of temperature changes between 35–38 °C.

Short-latency sensory evoked potentials are not affected by massive doses of sedatives. Thus, normal brain-stem evoked potentials in a patient with *isoelectric EEG, areflexy, and apnoe* suggests drug overdose. The presence of normal or near normal short-latency somatosensory evoked potentials in patients with a iatrogenic depressed EEG and motor reflex (such as it may be seen in high dose barbiturate or benzodiazepine overdose with cumulative effects) offers the likelihood of recovery. Poor quality outcome in patients after head injury or a raised intracranial pressure is preceded by incomplete recovery of later cortical components. They, however, return to normal in patients making satisfactory improvement.

Thus, SEPs per se give an early guide to the recovery of headinjured patients. Those with sustained absence of early evoked somatosensory potentials of the right and the left side will not recover. Those with marked asymmetry or severe abnormality of the peak, are at best capable of survival with severe damage or in a vegetative state. Those with minor short-latency asymmetries or morphological abnormalities of the wave eventually will recover fully even in spite of the early neurological deficits.

The combined use of the EEG and the EP gives greater accuracy to predict outcome than would be achieved with either test alone. Such a combination helps to overcome problems of interpretation of clinical data even in the presence of high levels of sedative drugs.

Additionally, in the isoelectric EEG after cardiac arrest with persistent cortical evoked potentials this can be a monitoring sign of later recovery when compared to cases with isoelectric EEG and an absent cortical EP. In the latter, neocortical death is likely if cortical components and/or short latency components show sustained absence [109].

Although brain-stem auditory evoked potentials (BAEP) are helpful in confirming brain death, some institutions believe the somatosensory evoked potentials (SEP) to be of greater clinical usefulness in the setting of brain death. This assumption is born out of the observation, that it is important to have an input signal suggesting the *arrival of the stimulus to the central nervous system.* Such a differentiation is not that easy when using brain-stem auditory evoked potentials [110]. No brain death patients do show absent late components.

Comatose patients with preserved brain-stem function, however, present a range of abnormalities which correlate with the neuropathological finding. It has, however, to be born in mind, that the causes of evoked potential abnormalities may also be within reversible toxic, or hyopthermic origin. Thus, it seems wise not to make a *vital decision* on the basis of *absent potentials* and emphasize waves, and potentials that are present.

21. Glossary

Abbreviations used in clinical electrophysiology

AC = alternating current
AE = activity edge
AEP = auditory evoked potential
AP = action potential
BAEP = brainstem auditory evoked potential
BAER = brainstem auditory evoked response
BEAM = brain electrical activity mapping
CBF = cerebral blood flow
CCT = central conduction time
CDSA = color density spectral array
CNG = cochlear nystagmogram
CSA = continuous spectral array
CV = conduction velocity
DC = direct current
DSA = density spectral array
ECochG = electrocochleogram
EEG = electroencephalogram or electroencephalograph
EMG = electromyogram
ENG = electronystagmogram
EOG = electrooculogram
EPSP = excitatory postsynaptic potential
EP = evoked potential, also Erb's point in illustrations
ERA = evoked response audiometry
ERG = electroretinogram
FED = flash emitting diodes
FEP = flash evoked potential
FFT = Fast Fourier Transform
FP = fossa poplitea
Hz = Hertz
IPSP = inhibitory postsynaptic potential
LED = light emitting diodes
LMN = left median nerve
LPTN = left posterior tibial nerve
MEP = motor evoked potential
MPF = median power frequency
PPF = peak power frequency
RMN = right median nerve
RPTN = right posterior tibial nerve
SAEP = short latency auditory evoked potential
SEP = somatosensory evoked potential
SpEP = spinal evoked potential
SSEP = short latency somatosensory evoked potential
SEF = spectral edge frequency
VEP = visual evoked potential
VER = visual evoked response

22. Bibliography

1. Blacher RS (1975): On awakening paralyzed during anesthesia, a syndrome of traumatic neurosis. *JAMA* 234:67.
2. Hilgenberg JC (1981) Intraoperative awareness during high dose fentanyl-oxygen anesthesia. *Anesthesiology* 54:341–343.
3. Greenberg RP, Griffith RL (1980): Neurophysiologic evaluation and monitoring of brain function. In: Cottrell JE, Turndorf H (eds), *Anesthesia and Neurosurgery,* pp. 401–420. St. Louis: Mosby.
4. Grundy BL (1983): Electrophysiologic monitoring EEG and evoked potentials. In: Newfield P, Cottrell JE (eds), *Manual of Neuroanesthesia*, pp. 28–59. Boston: Little Brown.
5. Levy WJ, Grundy BL, Smith NT (1984): Electroencephalography and evoked potentials. In: Saidman LJ, Smith NT (eds), *Monitoring in Anesthesia and Intensive Care,* pp. 227–267. London: Buttersworth.
6. Prior PF (1979): Monitoring cerebral function. In: *Long-term Recordings of Cerebral Electrical Activity.* Amsterdam: North Holland, Biomedical Press.
7. Baldock G, Walker, WG (1946): A new electronic analyser. *Electronic Engineer* 18:339.
8. Brechmert VL (1964): Current status of electroencephalography in the practice of clinical anesthesia. *Clin Anesth* 2:8.
9. Cooley JW, Turkey JW (1965): An algorithm for the machine calculation of complex Fourier series. *Mathemat Comput* 26:429.
10. Findeiss JC, Kien JA, Huse KOW, Linde HW (1969): Power spectral density of the electroencephalogram during halothane and cyclopropane anesthesia in man. *Anesth Analg* 48:108.
11. Ingram GS, Payne JP, Perry IR (1976): Computerized frequency analysis of the EEG during induction of anaesthesia. *Br J Anaesth* 48:275.
12. Jasper HH (1958): Report of the committee on methods of clinical examination in electroencephalography. *Electroencephalogr Clin Neurophysiol* 10:370.
13. Prior S, Faulconer A Jr, Bickford RG, Hunter RC (1953): Electroencephalographic pattern during anesthesia with cyclopropane – correlation concentrations in arterial blood. *Anesth Analg* 32:130.
14. Levy WJ (1984): Intraoperative EEG pattern: Implication for EEG monitoring. *Anesthesiology* 60:430.

15. Prior P (1979): *Monotoring Cerebral Function.* Philadelphia: Lipincott.
16. Creutzfeld O, Lux HD, Watanabe S (1966): Electrophysiology of cortical cells. In: Pupura DP, Yahr MD (eds), *The Thalamus*, pp. 209–230. New York: Columbia University Press.
17. Andersen P, Andersson SA, Lamo T (1976): Some factors involved in the thalamic control of spontaneous barbiturate spindles. *J Physiol* (London) 192:257–281.
18. Andersen P, Eccles JC, Sears TA (1964): The ventro-basal complex of the thalamus: Types of cells, their response and their functional organization. *J Physiol* (London) 174:370–399.
19. Levy WJ, Shapiro HM, Marnchak G, Meathe E (1980): Automated EEG processing for intraoperative monitoring. A comparison of techniques. *Anesthesiology* 53:223.
20. Bickford GR, Billinger T, Fleming NT, Stewart LE (1975): The Compressed Spectral Array: A pictorial EEG. Proc San Diego. *Biomedical Symp* 11:365.
21. Stoeckel H, Schwilden H, Lauwen PM, Schüttler J (1981): EEG indices for evaluation of depth of anaesthesia. *Br J Anaesth* 53:117.
22. Rampil IJ, Sasse FJ, Smith NT, Hoff GH, Flemming DC (1980): Spectral edge frequency: A new correlate of anesthetic depth. *Anesthesiology* 53:S12.
23. Fleming RA, Smith NT (1979): Density modulation: A technique for the display of three-variable data in patient monitoring. *Anesthesiology* 50: 543–546.
24. Cucchira RF, Sharborough FW, Messik JM (1979): An electroencephalographic filter processor as an indicator of ischemia during carotid endarterectomy. *Anesthesiology* 51:77–79.
25. Katzudsone RD (1981): EEG recording, electrode placement, and aspects of generator localization. In: Numez PL (ed), *Electrical Fields of the Brain. The Neurophysics of the EEG,* pp. 176–213. New York: Oxford University Press.
26. Demetrescu MC (1975): The aperiodic character of the electroencephalogram. *Physiologist* 18:189.
27. Gregory TK, Pettcus DC (1986): An electroencephalographic processing algorithm specifically intended for analysis of cerebral electrical activity. *J Clin Monit* 2:190–197.
28. Oppenheim AV, Scharfer RW (1975): *Digital Signal Processing*, pp. 532–571. Prentice Hall: Englewood Cliffs, N.J.
29. Stockard J, Bickford RG (1975): The neurophysiology of anesthesia. In: Gordon E (ed), *A Basis and Practice of Neuro-Anesthesia.* Amsterdam: Excerpta Media (Monograms in Anesthesiology, Vol. 2).
30. Clark DL, Hosik EC, Rosner BS (1971): Neuropathological effects of different anesthetics in unconscious man. *J Appl Physiol* 31:884.
31. Malkin M, Eisenberg D (1963): Correlation between clinical and electroencephalographic findings during the first stage of nitrous oxide anesthesia. *J Oral Surg* 21:16.
32. Clark DL, Rosner BS (1972): Neurophysiological effects of general anesthetics. *Anesthesiology* 38: 564.
33. Backman LE, Loefstroem B, Widen L (1962): Electroencephalography in halothane anesthesia. *Acta Anaesthesiol Scand* 8: 115.
34. Findeiss JC, Kien JA, Huse KOW, Linde HJW (1969): Power spectral density of the electroencephalogram during halothane and cyclopropane anesthesia in man. *Anesth Analg* 48: 1018.
35. Neigh JL, Garman JP, Harp JR (1971): The electroencephalographic pattern during anesthesia with Ethrane: Effects of depth of anesthesia, $PaCO_2$, and nitrous oxide.

Anesthesiology 35:482–487.
36. Homi J, Kouchigeri HM, Eckenhoff JB, Linde HW (1972): A new anesthetic agent-Forane (isoflurane): Preliminary observations in man. *Anesth Analg* 51:439.
37. Eger EI II, Stevens WC, Cromwell TH (1971): The electroencephalogram in man anesthetized with Forane. *Anesthestiology* 53:504–508.
38. Arfel G, Weiss J, Dubost CH (1962): Accidents convulsifs benign et hypothermie profonde. *Anesth Analg* (Paris) 9:661.
39. Freye E, Dehnen-Seipel H (1990): EEG-effects during rewarming in CABG using either enflurane or isoflurane anesthesia (in press).
40. Ingvar DH, Spolund B, Ardo A (1976): Correlation between dominant EEG frequency, cerebral oxygen uptake and blood flow. *Electroencephalogr Clin Neurophysiol* 41:268–276.
41. Burton GW (1964): Metabolic acidosis during profound hypothermia. *Anaesthesia* 19:118.
42. Darimont PC, Jenkins LC (1977): The influence of intravenous anaesthetics on enflurane-induced central nervous system seizure activity. *Can Anaesth Soc J* 24:42.
43. Neigh JL, Garman JK, Harp JR (1971): The electroencephalographic patterns during anesthesia with Ethrane. *Anesthesiology* 35:482.
44. Moruzzi G (1972): The sleep waking cycle. *Rev Physiol* 64:2–145.
45. Ghonneim MM, Yamada T (1977): Etomidate: A clinical and electroencephalographic comparison with thiopental. *Anesth Analg* 57:479.
46. Ferrer-Allado, Brechner VL, Dymond H (1973): Ketamine-induced electroconvulsive phenomena in the human limbic and thalamic regions. *Anesthesiology* 38:333.
47. Thompson GE (1972): Ketamine-induced convulsions. *Anesthesiology* 37:662.
48. Sebel PS, Bovill JG, Wauquier A, Rog P (1981): Effects of high dose fentanyl anesthesia on the electroencephalogram. *Anesthesiology* 55:203–211.
49. Klein SL, Klein VL, Eng D (1979): The electroencephalogram under fentanyl N_2O anesthesia. *Anesthesiology* 51:S3.
50. Bovill JG, Sebel S, Wauquier A, Rog P, Schuyt HC (1983): The influence of high dose alfentanil anaesthesia on the electroencephalogram: Correlation with plasma concentrations. *Br J Anaesth* 55:199s–209s.
51. Bovill JG, Sebel S, Wauquier A, Rog P (1982): Electroencephalographic effects of sufentanil anaesthesia in man. *Br J Anaesth* 54:45–52.
52. Freye E, Gupta BN (1979): A modified technique for the selective perfusion of the fourth cerebral ventricle in conscious dogs. *J Pharmacol Methods* 2:305–3145.
53. De Castro J, van de Water A, Wouters L, Xhonneux R, Renemann R, Kay B (1979): Comparative study of cardiovascular, neurological and metabolic side effects of eight narcotics in dogs. *Acta anaesth belg* 30:5–99.
54. Scott JL, Ponganis KV, Stanski DR (1985): EEG quantification of narcotic effect: The comparative pharmacodynamics of fentanyl and alfentanil. *Anesthesiology* 62:234–241.
55. Hug CL Jr (1985): Lipid solubility, pharmacokinetics and the EEG: Are you better off today than you were years ago. *Anesthesiology* 62: 221–225.
56. Howie MB, Kyre J, Hoffer L, MacDonald JS, Dzwonczyk R, Reitz J (1983): Pharmacokinetic and electroencephalographic confirmation of alfentanils short duration. *Anesthesiology* 59:A 318.
57. Saletu B, Grünberger J, Berner P, Koeppen D (1985): On difference between 1,5- and 1,4-benzodiazepines: Pharmaco EEG and psychometric studies with clobazepam and lorazepam. In: Hindmarch I, Stomer PD, Trimble MR (eds), *Human Pharmacology*

and Clinical Application, pp. 23–46. London: Royal Soc Med. (Int Cong Symp Series 71).

58. Hartung E, Freye E (1988): An open comparison of propofol and enflurane for prolonged abdominal operations. *Anaesthesia* 43:105.
59. Wodburn DM, Rollins LT, Gardner MD (1957): Effects of carbon dioxide on the brain excitability and electrolytes. *J Physiol* 129:79.
60. Rampil IJ, Holzer JA, Quest DO, Rosenbaum SH, Correll JW (1983): Prognostic value of computerized EEG analysis during carotid endarterectomy. *Anesth Analg* 62:186–192.
61. Smidt M, Sharborough FW, Anderson RE Michenfelder JD (1974): Cerebral blood flow measurements and electroencephalograms during carotid endarterectomy. *J Neurosurg* 41:310–320.
62. Trojabard W, Boysen G (1973): Relation between EEG regional cerebral blood flow and internal carotid artery endarterectomy. *Electroencephalogr Clin Neurophysiol* 34:61.
63. Niedermeyer E, Lopez de Silva F (1982): *Electroencephalography: Basic Principles, Clinical Application, and Related Field.* Baltimore: Urban & Schwarzenberg.
64. Silvay G, Mindich BP, Owitz S, Koffsky RM, Litwak RS (1983): Evaluation of a new cerebral function monitor during open-heart surgery. *MSJM* 546:44.
65. Sharborough FW, Messick JM, Smidt TM (1973): Correlation of continuous electroencephalograms with cerebral blood flow measurements during carotid endarterectomy. *Stroke* 4:674–683.
66. Ferguson GG (1982): Intra-operative monitoring and internal shunts: Are they necessary in carotid endarterectomy? *Stroke* 13:287.
67. Cohen ME, Olzowka JS, Subramamian S (1978): Electroencephalographic and neurologic correlates of deep hypothermia and circular arrest in infants. *Ann Thorac Surg* 23:238–244.
68. Branthwaite MA (1977): Prevention of neurological damage during open-heart surgery. *Thorax* 20:258–261.
69. Dawson GE (1954): A summation technique for the detection of small evoked potentials. *Electroencephalog Clin Neurophysiol* 6:65.
70. Grisolia JS, Widerholt WC (1980): Short latency somatosensory evoked potentials from radial, median, and ulnar nerve stimulation in man. *Electroencephalogr Clin Neurophysiol* 50:375.
71. Chiappa KH, Choi SK, Young RR (1980): Short-latency somatosensory evoked potentials following median nerve stimulation in patients with neurological lesions. In: Desmedt JE (ed), *Clinical Uses of Cerebral, Brainstem and Spinal Somatosensory Evoked Potentials*, pp. 264–281. Basel: S. Karger.
72. Grundy BL, Nash CL, Brown RH (1982): Deliberate hypotension for spinal fusion: Protective randomized study with evoked potential monitoring. *Can Anaesth Soc J* 29:452.
73. Vas GA, Gracco JB, Gracco RB (1981): Scalp recorded short latency cortical and subcortical somatosensory evoked potentials to peroneal nerve stimulation. *Electroencephalogr Clin Neurophysiol* 52:1.
74. Burke D, Skuse NF, Lethlean HK (1981): Cutaneous and muscle afferent components of the cerebral potential evoked by electrical stimulation of human peripheral nerves. *Electroencephalogr Clin Neurophysiol* 51:579.
75. Desmedt JE, Branco E (1980): Functional organization of far-field and cortical components of somatosensory evoked potentials in normal adults. In: Desmedt JE (ed),

Clinical Uses of Cerebral, Brainstem, and Spinal Somatosensory Evoked Potentials, pp. 27–50. Basel: S. Karger.

76. Stöhr M, Petruch F (1979): Somatosensory evoked potential following stimulation of the trigeminus nerve in man. *J Neurol* 220:95–98.
77. Chiappa KH, Choi SH, Young RR (1980): Short-latency somatosensory evoked potentials following median nerve stimulation in patients with neurological lesion. In Desmedt JE (ed), *Clinical Uses of Cerebral, Brainstem, and Spinal Somatosensory Evoked Potentials*, pp. 264–281. Basel: S. Karger.
78. Gomes T (1980): A study of peripheral, cervical, and cortical evoked potentials and afferent conduction time in the somatosensory pathway. *Electroencephalogr Clin Neurophysiol* 49:446.
79. Gracco RQ (1980): Scalp recorded potentials evoked by median nerve stimulation: subcortical potentials, travelling waves, and somatosensory potentials. In: Desmedt JE (ed), *Clinical Uses of Cerebral, Brainstem, and Spinal Somatosensory Evoked Potentials*, pp. 1–14. Basel: S. Karger.
80. Desmedt JE, Branco E (1980): Functional organization of far-field and cortical components of somatosensory evoked potentials in normal adults. In: Desmedt JE (ed), *Clinical Use of Cerebral, Brainstem, and Spinal Somatosensory Evoked Potentials*, pp. 27–50. Basel: S. Karger.
81. Grundy B (1985): Evoked potential monitoring, in: Blitt CD (ed), *Monitoring in Anesthesia and Critical Care Medicine*, pp. 345–411. New York: Churchill Livingstone.
82. Freye E, Buhl R, Schenk GK (1987): Somatosensory evoked potentials (SEPs) are affected differently by mu- and kappa-selective opioids in the dog. *Funct Neurol* 2:281.
83. Desmedt JE, Noel P (1980): Average cerebral evoked potentials in the evaluation of the lesions of the sensory nerves and of the central somatosensory pathway. In: Desmedt JE (ed), *New Developments in Electromyography and Clinical Neurophysiology*, pp. 271–351. Basel: S. Karger.
84. Benedetti G, Butler SA (1982) Event related potential correlates of analgesia: Comparison of fentanyl, acupuncture and nitrous oxide. *Pain* 14:327.
85. Chapmann CR, Jacobsen RC (1984): Assessment of analgesic states: Can evoked potentials play a role? In: Bromm B (ed), *Pain Measurements in Man*, pp. 233–256. Amsterdam: Elsevier.
86. Chapman CR, Chen ACN, Harkins SW (1979): Brain evoked potentials as correlates of laboratory pain: A review and perspective. In: Bonica JJ, Liebeskind JC, Albe-Fessard DC (eds) *Adv Pain Res Therapy*, pp. 791–803. New York: Raven Press.
87. Hume AL, Durking MA (1986): Central and spinal somatosensory conduction times during hypothermic cardiopulmonary bypass and some observations on the effects of fentanyl and isoflurance anesthesia. *Electroencephalogr Clin Neurophysiol* 65:46–58.
88. Desmedt JE, Cheon G (1980): Central somatosensory conduction in man: Neural generators and interpeak latencies of the far field components recorded from neck and right or left scalp and earlobe. *Electroencephalogr Clin Neurophysiol* 59:382–403.
89. Anziska B, Gracco RQ, Look AW, Feld AW (1978): Somatosensory far field potentials. Studies in patients with focal neurological disease. *Electroencephalogr Clin Neurophysiol* 1:51–68.
90. Allison T, Goff WR, Williamson PD, van Gilder JL (1978): On the neural origin of early components of the human somatosensory evoked potential. In: Desmedt JE (ed), *Clinical Use of Cerebral, Brainstem, and Spinal Somatosensory Evoked Potentials*,

pp. 51–68. Basel: S. Karger.

91. Dorfman LJ, Bosley TM (1979): Age related changes in peripheral and central nerve conduction in man. *Neurology* 29:38.
92. Jörg H, Meldorn HM, Podenski R (1980): Cerebral refractory period of somatosensory system: EEG and clinical findings before and after vascular surgery in cerebrovascular disease. *Expt Med* 526:181.
93. Robertson CS, Foltz R, Grossmann RG, Goodman JC (1986): Protection against experimental ischemic spinal cord injury. *J Neurosurg* 64:633–642.
94. Grundy BL (1982): Monitoring of sensory evoked potentials during neurosurgical operations. Methods and applications. *Neurosurg* 11:556.
95. Randzeus PA (1982): Intraoperative monitoring of evoked potentials. *Ann NY Acad Sci* 388:308.
96. Stöhr M (1988): Somatosensory evoked potentials-SEP. In: Maurer M, Lowitsch K, Stöhr M (eds), *Evozierte Potentiale.*, pp. 130–178. Stuttgart: Enke Verlag.
97. Hume AL, Cant BR, Shaw NA (1979): Central somatosensory conduction time in comatose patients. *Ann Neurol* 5:379.
98. Allen A, Starr A, Nudlenan K (1981): Assessment of sensory function in the operating room utilizing cerebral-evoked potentials. A study of fifty-six surgically anesthetized patients. *Clin Neurosurg* 28:457.
99. Riffel B, Stöhr M, Körner S (1984): Spinal and cortical evoked potentials following stimulation of the posterior tibial nerve in the diagnosis and localization of spinal cord diseases. *Electroencephalogr Clin Neurophysiol* 58:400.
100. Stöhr M, Dichgans J, Voigt K, Buettner UW (1983): The significance of somatosensory-evoked potentials for localization of unilateral lesions within the cerebral hemispheres. *J Neurol Sci* 61:49–63.
101. Mosko S, Pierce S, Holowach J, Sassni JF (1981): Normal brain stem auditory evoked potentials recorded in sleep apneics during wakening and as a function of arterial oxygen saturation during sleep. *Electroencephalogr Clin Neurophysiol* 51:477.
102. Little JR, Lesser RP, Lueders H, Furlan AJ (1983): Brain stem auditory evoked potentials in posterior circulation surgery. *Neurosurgery* 12:496.
103. Halliday HM (1983): *Evoked Potentials in Clinical Testing.* Edinburgh: Churchill Livingstone.
104. Chiappa KH (1983): *Evoked Potentials in Clinical Medicine.* New York: Raven Press.
105. Grundy BL (1985): Intraoperative application of evoked responses. In: Owen JH, Davis H (eds), *Evoked Potential Testing: Clinical Applications*, pp. 159–212. Orlando, FL: Grune & Stratton.
106. Cooper R (1980), *J Electrophysiol Technol* 9:22.
107. Boyd SG, Rothwell JC, Cowan JMA, Webb PJ, Morley T, Asselman P, Marsden CD (1986): A method of monitoring function in cortico-spinal pathways during scoliosis surgery with a note on motor conduction velocity. *J Neurol Neurosurg Psychiat* 49:251–257.
108. Rosenstein J, Wang AD, Symon L, Suzuki M (1985): Central conduction time and clinical grade of aneurysmal subarachnoid hemorrhage. *J Neurosurg* 62:25–30.
109. Trojaborg W, Jorgensen EO (1973): Evoked cortical responses in potentials with 'isoelectric' EEGs. *Electroencephalogr Clin Neurophysiol* 35:301–309.
110. Goldie WD, Chiappa KH, Young RR, Brooks EG (1981): Brainstem auditory and short-latency somatosensory evoked responses in brain death. *Neurology* 31:248–256.

Index of subjects

DEVELOPMENTS IN CRITICAL CARE MEDICINE AND ANESTHESIOLOGY

1. O. Prakash (ed.): *Applied Physiology in Clinical Respiratory Care.* 1982 ISBN 90-247-2662-X
2. M. G. McGeown: *Clinical Management of Electrolyte Disorders.* 1983 ISBN 0-89838-559-8
3. T. H. Stanley and W. C. Petty (eds.): *New Anesthetic Agents, Devices and Monitoring Techniques.* Annual Utah Postgraduate Course in Anesthesiology. 1983 ISBN 0-89838-566-0
4. P. A. Scheck, U. H. Sjöstrand and R. B. Smith (eds.): *Perspectives in High Frequency Ventilation.* 1983 ISBN 0-89838-571-7
5. O. Prakash (ed.): *Computing in Anesthesia and Intensive Care.* 1983 ISBN 0-89838-602-0
6. T. H. Stanley and W. C. Petty (eds.): *Anesthesia and the Cardiovascular System.* Annual Utah Postgraduate Course in Anesthesilogy. 1984 ISBN 0-89838-626-8
7. J. W. van Kleef, A. G. L. Burm and J. Spierdijk (eds.): *Current Concepts in Regional Anaesthesia.* 1984 ISBN 0-89838-644-6
8. O. Prakash (ed.): *Critical Care of the Child.* 1984 ISBN 0-89838-661-6
9. T. H. Stanley and W. C. Petty (eds.): *Anesthesiology: Today and Tomorrow.* Annual Utah Postgraduate Course in Anesthesiology. 1985 ISBN 0-89838-705-1
10. H. Rahn and O. Prakash (eds.): *Acid-base Regulation and Body Temperature.* 1985 ISBN 0-89838-708-6
11. T. H. Stanley and W. C. Petty (eds.): *Anesthesiology 1986.* Annual Utah Postgraduate Course in Anesthesiology. 1986 ISBN 0-89838-779-5
12. S. de Lange, P. J. Hennis and D. Kettler (eds.): *Cardiac Anaesthesia.* Problems and Innovations. 1986 ISBN 0-89838-794-9
13. N. P. de Bruijn and F. M. Clements: *Transesophageal Echocardiography.* With a contribution by R. Hill. 1987 ISBN 0-89838-821-X
14. G. B. Graybar and L. L. Bready (eds.): *Anesthesia for Renal Transplantation.* 1987 ISBN 0-89838-837-6
15. T. H. Stanley and W. C. Petty (eds.): *Anesthesia, the Heart and the Vascular System.* Annual Utah Postgraduate Course in Anesthesiology. 1987 ISBN 0-89838-851-1
16. D. Reis Miranda, A. Williams and Ph. Loirat (eds.): *Management of Intensive Care.* Guidelines for Better Use of Resources. 1990 ISBN 0-7923-0754-2
17. T. H. Stanley (ed.): *What's New in Anesthesiology.* Annual Utah Postgraduate Course in Anesthesiology. 1988 ISBN 0-89838-367-6
18. G. M. Woerlee: *Common Perioperative Problems and the Anaesthetist.* 1988 ISBN 0-89838-402-8
19. T. H. Stanley and R. J. Sperry (eds.): *Anesthesia and the Lung.* Annual Utah Postgraduate Course in Anesthesiology. 1989 ISBN 0-7923-0075-0
20. J. De Castro, J. Meynadier and M. Zenz: *Regional Opioid Analgesia.* Physiopharmacological Basis, Drugs, Equipment and Clinical Application. 1990 ISBN 0-7923-0162-5
21. J. F. Crul (ed.): *Legal Aspects of Anaesthesia.* 1989 ISBN 0-7923-0393-8
22. E. Freye: *Cerebral Monitoring in the Operating Room and the Intensive Care Unit.* 1990 ISBN 0-7923-0349-X
23. T. H. Stanley and R. J. Sperry (eds.): *Anesthesiology and the Heart.* Annual Utah Postgraduate Course in Anesthesiology. 1990 ISBN 0-7923-0634-1

KLUWER ACADEMIC PUBLISHERS – DORDRECHT / BOSTON / LONDON

GPSR Compliance

The European Union's (EU) General Product Safety Regulation (GPSR) is a set of rules that requires consumer products to be safe and our obligations to ensure this.

If you have any concerns about our products, you can contact us on ProductSafety@springernature.com

In case Publisher is established outside the EU, the EU authorized representative is:

Springer Nature Customer Service Center GmbH
Europaplatz 3
69115 Heidelberg, Germany

Batch number: 10607214

Printed by Printforce, the Netherlands